DEEP, DARK AND DANGEROUS

Deep, Dark and Dangerous

The Story of British Columbia's
World-Class Undersea Tech Industry

VICKIE JENSEN

HARBOUR
PUBLISHING

Harbour Publishing Co. Ltd.

P.O. Box 219, Madeira Park, BC, V0N 2H0

www.harbourpublishing.com

Edited by Lynn Van Luven
Indexed by Chandan Singh
Text and dust jacket design by Shed Simas / Onça Design
Printed and bound in Canada

Harbour Publishing acknowledges the support of the Canada Council for the Arts, the Government of Canada, and the Province of British Columbia through the BC Arts Council.

Library and Archives Canada Cataloguing in Publication

Title: Deep, dark and dangerous : the story of British Columbia's world-class undersea tech industry / Vickie Jensen.
Names: Jensen, Vickie, 1946- author.
Description: Includes index.
Identifiers: Canadiana (print) 2021027106X | Canadiana (ebook) 20210271205 | ISBN 9781550179200 (hardcover) | ISBN 9781550179217 (EPUB)
Subjects: LCSH: Ocean engineering industry—British Columbia. | LCSH: Ocean engineering—Technological innovations—British Columbia. | LCSH: Submersibles—Technological innovations—British Columbia.
Classification: LCC HD9999.O343 C25 2021 | DDC 338.4/7620416209711—dc23

Contents

Preface

When people ask me what I'm writing, their eyebrows invariably shoot up when I reply "BC's subsea history." Yes, it's an unusual subject, especially since I spent my childhood in the cornfields of the US Midwest. But I vividly remember the rare occasions when my father told me about his early Navy days as a helmet diver and how he'd almost drowned when his dive attendant hadn't immediately hauled him to the surface after his helmet flooded. Like so many others of my generation, I also was captivated by the early television show *Sea Hunt*, in which Lloyd Bridges portrayed Mike Nelson, a former US Navy frogman. For me, the underwater world has always been both dangerous and exciting.

As a child, I dreamed of a pair of magic glasses that would let me see underwater. That fantasy was no doubt reinforced the summer I dropped my mother's diamond ring off a lakeside dock. My father shouted for everyone in the water to stay absolutely still, then grabbed a diving mask from a nearby kid, dove down and located the ring. Magic glasses, indeed!

Decades later, having moved to the coast of British Columbia, I landed the dream job of editing *Westcoast Mariner* magazine. Every month for four years, I went out on different types of coastal workboat—tugs, charter boats, dredges, Coast Guard launches, pile drivers, pilot boats—in order to write about the vessel, its crew and their jobs. I am pretty sure that I was the only woman whose "Dress for Success" wardrobe included a pair of gumboots in her bottom desk drawer.

One day, a reader named Harry Bohm contacted me. "You write about everything happening *on* the water. What about covering what's happening *under*water?" At the time, Harry had traded working on tugs for managing

Simon Fraser University's Underwater Research Lab. He introduced me to the world of submersibles and unmanned, robotic craft like ROVs (Remotely Operated Vehicles with their long tether) and AUVs (tetherless Autonomous Underwater Vehicles). I also began meeting some of the remarkable folks who were working, inventing and researching underwater. He was right—there was an entire subsea world to learn and write about!

A year later, Harry was back. "OK, now I want you to write the book I wish I'd had as a kid." He told me about being fascinated by Jacques Cousteau's subsea habitats and, of course, the TV show *Sea Hunt*. In secondary school he'd built a miniature Cousteau-habitat for his science fair project and actually raised a mouse underwater. Now, he was working on the concept of small hardware-store-technology ROVs that even kids could build and operate. Intrigued, I said, "Yes!"

A year later, we published *Build Your Own Underwater Robot and Other Wet Projects*, including a glowing recommendation from adventure-author Clive Cussler for the back cover. The coincidence was that the do-it-yourself underwater robot projects detailed in that book were simply a high-tech version of the magic glasses I'd dreamed of as a child.

Several years later, Harry and I teamed up with Dr. Steve Moore, an underwater-robot-building zoologist and professor at California State University Monterey Bay (CSUMB). The three of us invested years writing the landmark textbook *Underwater Robotics: Science, Design & Fabrication*, published by the Marine Advanced Technology Education Center (MATE) in Monterey, CA. It provides an important resource for the "underwater robots in the classroom movement," which is fostered by MATE's annual student ROV competition. All these opportunities, in combination with hundreds of hours of recent interviews, have shaped my fascination with the subsea world.

From the outset I knew that the title of this book had to be *Deep, Dark and Dangerous*. Those three words perfectly describe the risk and conditions of underwater work. Daunting as they are, those words never defeated the determination of BC's underwater pioneers. They routinely took on risks of all kinds, testing their bodies, their inventions and their financial resources. They refused to believe something was impossible—it just hadn't been done yet.

Early on, I was amazed to realize that hardly anyone outside the subsea industry knew anything about these risk-takers or their accomplishments.

As well, it was little known that, beginning in the mid-1960s, BC's undersea tech industry developed a world-class reputation. That legacy is still acknowledged globally.

British Columbia's undersea story presents a vital chapter of largely unrecognized Canadian history. Determined to change that glaring omission, I proposed the idea to Harbour Publishing. Howard White immediately agreed. The result is *Deep, Dark and Dangerous: The Story of British Columbia's World-class Undersea Tech Industry.*

CHAPTER 1

Al Trice: The challenge of working underwater

1950s and '60s

Almost without exception, British Columbia's subsea visionaries started off as divers. Early on, they gravitated to the water, read Jacques Cousteau's books and watched the TV drama series *Sea Hunt*, which ran in the late 1950s and early '60s. Most got their scuba certification as soon as they could, often in their mid-teens. Many then graduated to the heavy helmets and suits of commercial diving, a work-world where quirky weather, brutal conditions and rigid timelines were the norm. They learned by testing themselves and their equipment in the water. Often working far from the availability of supplies or spare parts, they "made do," took chances, and innovated on the fly. The wisdom they gained led to pragmatic revisions, ideas for future inventions and a critical awareness of the employer's requirements.

In his unpublished manuscript "Underwater Mobility in Canada, 1800 to 2007: A historical assessment," Dr. David McGee notes that revolutionary change happened in commercial diving after 1945: "In the 1950s, typical [commercial diving] depths were still less than 100 feet. By 1962, drilling off the coast of California called for dives of 200 to 250 feet. By 1965, there were contracts calling for 500 and 600 feet in various parts of the world. By the end of the decade, partly due to the development of North Sea oil, 900 feet was common, and 1,000 feet on the horizon. By 1980, commercial divers would reach an astounding 2,000 feet." These first two chapters introduce BC diving legends Allan "Al" Trice and Phil Nuytten. Their exploits also highlight the challenges, conditions and risks of working underwater—all part of that astounding revolution.

A winter dive on the Fraser

It's a blustery January day in the late 1950s, with an icy wind targeting a diving scow on BC's Fraser River. Al Trice is suited up in hardhat diving gear with its trademark brass helmet, air hose, a waterproofed canvas suit, diving knife, adjustable weights to counteract buoyancy, heavy metal shoes and a diver's telephone. This "standard diving dress," as it's known, totals around 190 lbs (85 kg)—more than Al himself weighs! To help insulate against the cold January waters, he starts with several suits of full-length woollen underwear, three pairs of woollen socks and a woollen balaclava head cover. Early on, there were no suitable gloves, so Al and others worked bare handed. On winter jobs in ice-covered lakes in the BC Interior or Prairies hypothermia is a danger, so often two divers work a single job, trading diving and tendering positions every two hours.

The surface member of a two-man team is the dive tender. Having already checked equipment and topside air supply and helped dress the diver, the tender now locks the heavy brass helmet onto Al's diving suit. Then he does a final check of the air supply as well as the telephone communication wire taped to the diver's air hose that will allow talk between surface and murky bottom. One of the telephone speakers hangs on a post for the dive tender. A second speaker is conveniently inside the diver's helmet. On jobs such as this, involving a crane, there's a third speaker stretching up to the crane operator, high above the scow, so he can also follow the diver's progress. On the surface, the dive tender monitors communications and the compressor that pumps vital air down to the diver.

Al hefts the pneumatic chainsaw he'll use to cut off the wooden pilings of an old wharf. It will work well at these shallow depths, but he knows an air-supplied saw operates with less power as jobs get deeper. Carefully, he makes his way down the scow's wooden ladder into the Fraser's raw January waters, reminding himself that this is a relatively easy commercial dive job and it pays decently.

On the surface, two bridgemen working in a skiff secure a wire sling that runs from the crane to the first piling. The crane operator takes up the slack and waits for word from the diver. In winter, Al spots the old pilings when he's 2 to 3 feet away. In summer, the silt in the Fraser turns the water to mud, and he would have to grope to find them, relying on his "10 eyes," as divers call their fingers. Al starts up his chainsaw and begins on the first piling. Once he's almost completed cutting through the thick base, he alerts the surface crew and the crane driver in particular. Both operator and diver know that once cut loose, a piling will rocket up to the surface, often turning end over end in the air. Without the restraining wire for control, the hefty piling could crash back into the water and injure the diver. Very buoyant pilings can even shoot high enough out of the water to damage the crane. The last step is for the bridge-men in the skiff to secure the valuable piling for re-use, then attach the sling to the next piling and alert the diver.

In January, a couple of hours underwater feels like a long, numbing shift. On the work scow, a diving shelter and oil stove afford welcome, if limited, protection from the weather for the tender and a place for the returning diver to warm up. Hot coffee is always on. In the early days, a kettle of hot water was also kept boiling because when a diver re-surfaced in below-zero weather, his helmet was often frozen onto the collar of his diving suit. Hot water poured over the helmet quickly melted the ice.

In summertime, Al's workday starts early, often by 6 or 7 a.m., like any construction job. But there the similarity ends, since it takes a hardhat diver about half an hour to get into his heavy standard diving dress. Once suited up, he heads underwater for three to three and a half hours before coming back to the surface for a half-hour lunch. Then it's another half hour to get dressed again for the afternoon shift. Generally, commercial divers spend six to seven hours in the water if cold is not a factor. Depending on the job, they may work seven days a week, especially if they're part of a construction crew out in the bush, living in a nearby logging, fisheries or mining camp.

This job in the Fraser is comparatively easy. Cutting wooden pilings is far less complex than using hydraulic compression drills to bore holes in order to plant explosives, hefting a cutting torch to slice through steel underwater, or building coffer dams for bridge construction. Big jobs like the Port Mann Bridge, constructed in 2009–12, often run a round-the-clock schedule, with three diving crews working eight-hour shifts.

Al recalls that most kids entering the job market back in the '50s and '60s were shaped by two major world events—the Depression of the '30s and the Second World War in the '40s. Al joined Army Cadets in Grade 7 and mastered Morse Code. He also learned how to strip a Sten gun and a 30-calibre Browning machine gun, but Cadets weren't allowed to touch hand grenades. Magee Secondary School in Vancouver, where Al grew up, had an indoor shooting range complete with Enfield rifles, which he and a buddy excelled at. So, it might seem that after the war a comfortable office job with a regular paycheque would have been every kid's goal. Not so. In fact, commercial diving is about as far from safe, predictable work as it's possible to get.

Working on the water

Al's dad built small boats, taught his son to build dinghies and helped Al sneak out of the house on solo explorations with his own 15-ft (4.5 m) sailboat. He ventured all over Howe Sound, across the Gulf of Georgia and down to Victoria. Earning one's keep was an important family value, so Al cut lawns and delivered morning and evening newspapers as a kid; later he worked in a butcher shop after school and built dinghies for cash. As a wiry 15-year-old, Al Trice already knew boats and the water. But he also loved flying: he rebuilt a full-sized glider and got his B licence after five solo flights. The same year he built a 23-ft (7-m) Star sailboat and began racing.

Although Al dreamed of becoming an aeronautical engineer, boats and the water won out. After secondary school he left home to spend a couple of seasons on a whaling station at Coal Harbour in northern Vancouver Island and a season salmon-trolling on the *Panda II*. He returned home, got married, and spent a year running a company towboat for Harbour Services in Vancouver harbour and Georgia Strait—a respectable marine education for a young man growing up in BC. At the age of 23 Al decided to learn more about wooden shipbuilding and apprenticed to Star Shipyard on the Fraser River, the biggest wooden shipyard on the coast at that time. It was 1952; his starting pay was 50 cents an hour.

Al started out building boats and got tagged with the nickname "Blondie." But he soon got fed up with Star Shipyards' archaic methods of utilizing only manual labour. He moved from constructing boats to lofting, the stage in between designing a boat and building it. "In those days, you lofted a boat by drawing it out full size. My first boat was about 65–70 feet long. That's how

I learned." Then Al was moved into the shipyard's joiner shop, where all the furniture and trim for a ship were fabricated. Built-in bunks, tables, windows and doors involved fancier, more exacting finish work.

"Frank Milne was a typical little old English joiner," Al recalls. "He ran the shop and was quite fussy, so I learned a lot from him. The first time I left one of my wood planes on the edge of the bench, he came by and knocked it off. Of course, it fell on the floor and the handle broke. 'Oh, you shouldn't have left it on the edge like that,' he told me." Al would adopt that tough-love teaching technique years later. When the head joiner took ill, Al was left to run the shop by himself.

The idea of diving underwater with scuba gear quickly captivated Al Trice. His brother-in-law, a commander in the Canadian Navy, had learned to scuba dive with Jacques Cousteau, co-inventor of the aqualung. Home on leave, he mentioned to Al that a guy was going to give a demonstration of "frog gear" at Crystal Pool in English Bay. "I went and was totally fascinated," Al remembers. "The guy was Keith Carter, an ex-British Navy frogman, and we had quite a talk afterwards. He was using a closed-circuit system, breathing pure oxygen with a full-face mask, bag and scrubber, so there were no bubbles. After seeing an ad in *Popular Science* for an aqualung with double tanks, Keith had also sent away for that, so I got to try both the closed-circuit unit and the aqualung."

Al started scuba diving with a closed-circuit oxygen system. "We used to bounce dive to 60, 70 feet, just for the hell of it. But just going down and coming right back up, you'd get terrible headaches. And those systems can cause all kinds of problems, convulsions, and even drowning. That's why Jacques Cousteau had gotten together with Émile Gagnon in 1943 and developed the aqualung. So I said, 'Geez, I gotta have an aqualung.'"

In the 1950s, scuba gear was scarce in BC, so Al and Keith jumped in Trice's MG TC roadster and took off for California and bought a set. "Once we both had this magic air-breathing aqualung, we instantly started roaring up and down 250 feet underwater, just for the hell of it and to see how well this thing would work. Fortunately, neither of us was too affected with nitrogen narcosis, which is like downing a couple of martinis at depth. Some people suffer badly from it, some don't."

Keith had already started a diving company with his brother, but the brother bailed out, so Al took his place. It was called Carter Brothers. "We kept the name because they already had a bunch of business cards printed up. So I left the shipyard to become a diver. That was in May of 1953, the same month that Hillary conquered Everest."

Making a living was a bit difficult at first because nobody knew much about scuba diving. "We took turns diving off the dock, mostly picking up lost freight that had slipped out of a sling. Also, a couple of deep-sea ships lost their anchors, and since they couldn't sail without one, it was a big deal to recover one. We'd charge $40 for half a day of picking up normal stuff, but for an anchor, we charged $600. Once we even retrieved an heirloom diamond ring and another time a set of dentures! Before the police forces had their own diver, recovering the bodies of drowning victims was another callout.

"We used to chug around the harbour in a little 12-foot clinker boat. Burrard Drydock, one of the large shipyards in North Vancouver, had an in-house hardhat diver doing work on their marine railway—that's the submerged railway used for hauling ships out of the water and onto a cradle for cleaning or repairs. We tied our boat up and decided to go underwater and take a look. The hardhat diver's support crew spotted our trail of bubbles and wondered, 'What the hell is this?' We came up, waved at them, then went back underwater. Well, word got around pretty quick. Management was going to rebuild their 600-ton marine railway, so wanted to talk to us. We told them both of us would be diving, and it would cost $160 a day. That was fine with them. So, for those three months, we made lots of money!"

In those days, all underwater work, both helmet and scuba, was generally done by single divers. Equipment failure was not an option from the standpoint of both one's own body and the customer's wrath. Al got so he could strip and repair an aqualung regulator in his sleep.

When Keith decided he'd had enough of diving, Al continued on his own. During the three months of work at Burrard Drydock, he'd gotten to know George Hazelton and Al Black, respectively the tender and hardhat diver there. "Blackie was Canadian and an enterprising guy who had ended up doing mine disposal in the Royal Navy. Eventually, he decided to leave the shipyard and go salvage logs for Powell River Company. George was going, too. I already knew about log salvage because Keith Carter and I had done that in Cowichan Lake on Vancouver Island, so they asked me to join in. In June of 1956, the three of us formed Universal Diving Company and headed up to Powell River. Once there, we needed a small boat we could use to set wire rope chokers on sunken logs, so I designed and built a 16-foot skiff in five days. Up there, sometimes we were divers and other times loggers in hard hats and caulk boots. George and I also learned hardhat diving from Blackie, and how to use explosives, too."

After six months of log salvage, the trio came back to Vancouver and decided to join the underwater construction union. The pile-driving union governed all the waterfront for bridges and dams and all underwater construction work. Al recalls: "In order to work on any of it, you had to belong to the union, but it was a closed shop. They had one diver and that was it. But various contractors badgered the union to hire more divers, so a bunch of us were let in. I was the only dual diver in the union. Often, I'd be diving hardhat in the morning and in scuba gear in the afternoon." Contractors soon realized that scuba gear was faster for surveying and similar jobs. Al quickly became one of the top commercial divers on the West Coast.

"We worked a lot of underwater construction jobs as well as contracting out to mining companies. We also did underwater welding, cutting, explosives inspection, repairs, bridge work, drilling and salvage. I also taught scuba-diving courses, got into underwater photography, and spent a lot of time designing, fabricating and improving the equipment we used underwater." Al made cutting torches, jetting nozzles and various custom valves in his dad's small basement machine shop. His father was a pattern maker for a foundry and helped out with the castings. Al says, "In the fall of 1969 I tallied up my time and discovered I'd spent 16½ years as a commercial diver and logged over 12,000 hours underwater."

There's got to be a better way

In 1964, Al Trice and his new diving partner Don Sorte were part of a six-diver contract that McKenzie Barge and Derrick secured with the Canadian government. The job was a salvage survey of a sunken oil scow, *Barge 10*, that had gone down in 260–330 ft (80–100 m) of water near tiny Pasley Island, at the entrance to Howe Sound. There were two scuba teams: Al and Don, and Ralf Somerville with Denis Tulin. George Hazelton worked as a single

Diving partners Don Sorte and Al Trice in standard diving dress, circa 1962. The difficulty and danger of ever-deeper dive jobs sparked the idea of building a small manned submersible. Personal photo collection of Al Trice

hardhat diver, as did Joe Hartle. Diving on such a deep job meant lengthy decompression times.

The Workmen's Compensation Board (now Worksafe BC) became involved because of the diving depth and assigned a doctor to monitor diver health. Although not a diver himself, he was interested in the effects of pressure breathing. "We were doing 25 minutes bottom time, five days per week," Al recalls, "but the doc cut that back to 14 minutes and four days per week."

The hardhat divers decompressed in the water, taking the mandated time coming back to the surface. They could easily vary their buoyancy by putting more or less air in their suits by using air valves and exhaust valves in their helmets. Deep diving also involved significant decompression times for the two scuba teams, but they handled it differently, utilizing a walk-in, two-compartment recompression chamber and compressor on board the work barge.

"The scuba divers would come up to 40 metres for one minute," Al explains. "After some intermediate stops to 9 metres for maybe 20 minutes or so, we'd hit the surface and zip out of our gear in the first chamber. We had three or four minutes to do that without suffering any bad effects. Then we'd move into the second chamber, and they'd pressurize it to 30 feet; we'd put on oxygen masks, climb into sleeping bags and go to sleep for two hours while being decompressed. Breathing pure oxygen cut our decompression time in half. Usually we'd go into a deep sleep because you burn a lot of energy working at those depths. To prevent the dangerous build-up of oxygen in the chamber, a high volume of air was piped through. Once we had moved into the second chamber and the door was secured, the first chamber was de-pressured so our suits could be cleaned of bunker oil. After the two hours of decompression were up, we traded jobs, working as tenders for the other divers, since we were only allowed one dive per day. Twelve-hour workdays were the norm in order to accomplish anything. It was one of the most difficult jobs ever."

Al's experience on the *Barge 10* salvage was not unique. Over the years, work-related danger increased as the depths for both hardhat and scuba diving jobs plummeted and decompression times lengthened. Deeper diving mandated longer decompression times—on the way back to the surface a diver either had to hang onto a weighted shot line at various depths for the amount of time necessary to clear his blood of excess nitrogen bubbles or come back to the surface and immediately spend hours in a decompression chamber. Divers who resurfaced directly or without sufficient decompression time could suffer the often crippling and sometimes fatal "bends." Most professional divers knew of co-workers who had suffered debilitating injuries—or who died on the job.

The dive tables that divers relied on to prevent decompression sickness were US Navy modifications of Scotsman John Scott Haldane's experiments on goats in the early 1900s. "But even those Navy tables were based on the results for divers in decompression tanks, not doing real work," veteran diver Phil Nuytten explains. "We knew those tables were inaccurate generalizations because a lot of people were getting 'hit' [by the bends] and hurt. For commercial divers, it was not unusual to get the bends several times, even in a week of work. It was pretty hairy stuff. So we started keeping careful notes on depths and times of dives, as well as water temperatures and altitude. Slowly we developed tables that were more relevant to specific diving conditions, whether that meant very deep dives, very long dives or in conditions such as Arctic waters." Hardhat and scuba equipment improved, as well. And slowly the expectations to "get the job done no matter what" evolved to include more attention to diver safety.

The difficulty of the *Barge 10* salvage experience changed Al's thinking—and eventually the future of subsea robotics in BC. "After that job, Don and I decided that there had to be a better way of deep diving in cold water. We had heard about mixed-gas diving using a combination of oxygen, helium and nitrogen for deep dives and thought that might be an option." At that time, the only place there was much mixed-gas diving, going down to 400 ft (120 m), was in Santa Barbara. So Don and Al hustled off to California and talked with divers there. But in the end, they decided against mixed gas because the decompression times were still unbelievably long.

"Instead, Don and I came to the conclusion that what we really needed was a small submarine, some kind of a vehicle that would take us down to 2,500 feet because that was how deep the fjords were in BC. We started off with a visit to Patterson Boiler Works down in the False Creek area. They'd built us a decompression chamber, but nixed building a submarine. So we hustled off down to California again, this time to get a submarine, because you could buy anything in California. But guess what? We couldn't find one!

"All these aerospace guys at Lockheed and other companies in California were trying their hand at building underwater vehicles since the outer space action had started to wind down. They called us the T-shirt boys because that's what we always wore. They saw us as hicks, just harmless Canadians from up north. We hardly posed any threat, so they proudly showed us their innermost secrets, things like the early moon buggy." Looking at their projects, Al quickly realized, "These things are never going to work. They're way too expensive and they're taking way too long to build."

One man in California had a little two-man submarine built by Chicago Bridge and Iron. It was rated to a 1,000-ft depth (305-m)—not as deep as they wanted, but it would get them started. "Don knew a customs broker, and it's a good thing that we talked to him first, because he reported that dear old Canadian Customs wanted to know what size the sub was and how many torpedo tubes it had! Well, we knew right then and there that it would take years to get this thing cleared to bring into Canada. So, we thought, shit, we're going to have to build the bloody thing *in* Canada."

In the meantime, Al and Don met with Gary Keffler, a dive shop owner in Seattle, and told him about their grand idea. Keffler thought he knew someone, an inventive Boeing machinist by the name of Mack Thomson, who might be interested in building a small submarine. "So we made arrangements to meet Mack in a café. And on a paper napkin the three of us planned a $20,000 submersible we'd build in three months!

"It would be the first one for British Columbia—heck, in all of Canada! We were determined to make it happen. Little did we know then that it would take 26 months and cost $100,000."

CHAPTER 2

Phil Nuytten: Pushing the limits

Late 1950s and '60s

As a 16-year-old, Phil Nuytten was a good-looking blond-haired kid with a ready smile. He had a hefty dose of self-determination, having dropped out of high school to open Vancouver's first dive shop, located on West 4th Street. Later he joined up with two older partners, moving the store to larger premises on Victoria Drive where they kept an illegal tugboat radio that enabled Phil to get the jump on paying salvage work. One June morning in 1958 is especially seared in his memory. He was out back, getting into his wetsuit pants in preparation for a small salvage job, when Bill Bamford, one of his partners, came running out of the shop shouting, "You gotta hear this!"

Voices were screaming on the tugboat radio: "It's coming down! Jesus, the whole bridge is coming down!" At first, Phil assumed the Lions Gate Bridge was in trouble, but soon realized it was the new Second Narrows Bridge, still under construction. As they listened to the broadcast, two entire spans of the bridge collapsed into Burrard Inlet. Phil grabbed his dive gear, jumped into his partner's truck and shouted for him to call for a police escort. The police caught up with Phil, sirens blaring, and they arrived at a scene of chaos, destruction and death. The collapse of the bridge had hurled mangled steel, heavy equipment and some 79 workers into the water below.

Phil still recalls the horror of that scene. "Eight or ten bodies had already been pulled out of the water onto the dock and were lined up in a row. I was just 16 years old and had never even seen a dead body before. All of a sudden, I was surrounded by them."

Phil was the first rescue diver in the water, with its rushing current. He spotted a struggling ironworker trapped inside a crane with his legs pinned. The man was already in water up to his chest, with the tide coming in fast.

Sixteen-year-old Phil Nuytten, suddenly turned rescue diver, was soon joined by fellow divers Don Sorte and "Scratch" McDonald. Swimming was out of the question in the 5 kph rip tide, so Phil Nuytten clung to downed steel, awaiting a boat tow. Photo by Brian Kent, *The Vancouver Sun*, p. 153 *Tragedy at Second Narrows: The Story of the Ironworkers Memorial Bridge* by Eric Jamieson

When divers Don Sorte and Deloye "Scratch" McDonald arrived, Phil dove down to see if he could free the man's legs, even taking off his tank to try to squeeze into the space. Nothing worked. "When I came back to the surface, Don just signalled me with his hand across his throat. The fast, incoming tide had already covered the poor man's head, and all I could see was his hair waving in the current.

"The scene was crazy everywhere. I saw a couple of ironworkers seemingly standing upright on the bottom. That was bizarre. They were all wearing their required old-time life jackets, but none of the flotation was powerful enough to counteract the weight of the heavy tool belt that everyone had on. So the belts kept these guys anchored on the bottom, but the life jackets made them stand up, bent forward in the current, with their arms sort of waving in the water. I can't forget that image. Every year, I get invited to attend a memorial service for all those guys and to talk about what it was like on that horrible day."

In all, 18 ironworkers died, with 20 more seriously injured. A Vancouver newspaper covering the event would note a row of corpses covered by sheets of canvas, all with their heavy work boots sticking out. A subsequent inquiry would conclude that the collapse was the result of an engineering calculation error that was eventually determined to be negligence on the part of the Dominion Bridge Company and "lack of care" on the part of Swan Wooster consulting engineers. The failure of the Second Narrows Bridge, later renamed the Ironworkers Memorial Second Narrows Crossing, was Phil Nuytten's first experience with maritime tragedy. It wouldn't be his last.

Totem poles, canoes and books

Strangely, it was totem poles, not scuba gear, that first intrigued Phil Nuytten. His folks owned a successful Vancouver restaurant on the corner of Chilco and Robson streets, so as a boy, nearby Stanley Park became his backyard. "At Lumberman's Arch there was a covered shed with chicken-wire walls that housed old Indian canoes. It was easy for a small kid to lift the wire and crawl under. I'd sit in one of the canoes and imagine all sorts of voyages and races (which I always won)." There were also a number of standing totem poles on the grassy area beside the canoe shed, and Phil used to climb up on them. "According to my mother, I told her that one day I was going to make one of those." Not surprisingly, by the time he was 11, Phil was learning to carve totem poles as a student of famous Kwakwaka'wakw artist Ellen Neel.

Phil's legal name is René Theophile Nuytten, but he has used "Phil" since he was a child, since his father's name was also René and his grandfather, a well-known character in St. Boniface, Manitoba, was Theo. This side of the family has a strong Metis heritage, and as a kid, Phil remembers hearing people describe his family as "dirty Indians." But he was affected positively by his dad telling him some of his own maternal grandfather's stories of bison hunts or trapping in the forest at Le Pas, Manitoba, and of the Metis people in St. Boniface and the Red River Valley. Phil also learned that his great-great uncle was a significant part of the Riel Rebellion. Growing up in BC, young Phil became fascinated with Northwest Coast Indigenous culture. As an adult, he's been accepted as a potlatch chief of the Mamalilikulla people. He's been given several names at various potlatches—Tlaxwsam ("Red Snapper"), Max'wa ("Always gives away") and Pu'kli?di ("Guests never leave hungry")—and adopted into several high-status Kwakwaka'wakw families. It may seem like a strange sideline to subsea work, but it's a crucial aspect of his complex identity.

Today, Phil Nuytten is most widely known as BC diving's "Renaissance man," an apt accolade considering he's not only a diver but an inventor, a tech manufacturer, a businessman, an adventurer, underwater explorer, author, owner of *Diver* magazine, a songwriter, a collector and a Northwest Coast Indigenous carver. Phil has a slightly different take on his wide range of interests. "The truth is, I have a short attention span so I like to have a number of projects on the go at once. That way, if I get tired of doing something or I've been at it for so long that I'm not doing a good job any more, then I jump to something else and come back fresh. As a little kid I always wanted to do what I wanted to do and didn't like anyone telling me I couldn't do it. Hmmm, not much has changed."

Phil read everything as a kid—even the back of Corn Flakes boxes. "Every Saturday morning, I would go to the book area at the Hudson's Bay department store. There was a wonderful lady in their book department by the name of Ella, I think. She used to let me read books for two or three hours every Saturday, but first I had to show her that I had washed my hands." There he discovered Hans Hass's book *Diving to Adventure: The Daredevil*

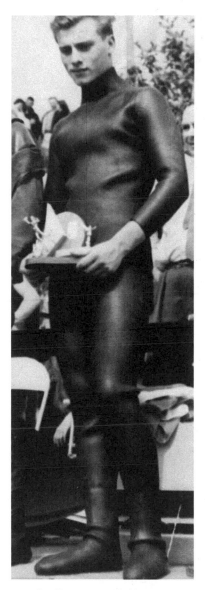

As a skindiver, young Phil Nuytten won trophies in breath-hold diving and spear-fishing competitions.
Personal collection of Phil Nuytten

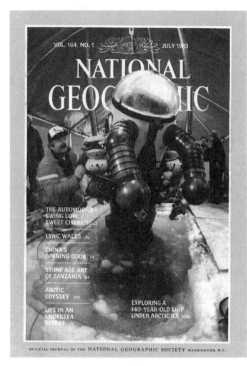

VOL. 164, NO. 1   JULY 1983

NATIONAL GEOGRAPHIC

THE AUTOMOBILE:
SWING LOW
SWEET CHARIOT! 2

LYRIC WALES 36

CHINA'S
OPENING DOOR 64

STONE AGE ART
OF TANZANIA 85

ARCTIC
ODYSSEY 100

LIFE IN AN
UNDERSEA
DESERT 129

EXPLORING A
140-YEAR-OLD SHIP
UNDER ARCTIC ICE 104A

OFFICIAL JOURNAL OF THE NATIONAL GEOGRAPHIC SOCIETY WASHINGTON, D.C.

Phil Nuytten, shown in a *WASP* diving suit, worked with Dr. Joe MacInnis, who led the expedition to reach the wreck of the *Breadalbane* in Arctic waters. Personal collection of Phil Nuytten

Story of Hunters Under the Sea and concluded that Jules Verne was a couch potato by comparison. As he got older, Phil persuaded his schoolmates to sign up for library cards; by using their extra cards, he could borrow six books at a time. Today his North Vancouver office is filled with 500–600 books, with another 1,500 at home. He's on his sixth Kindle and keeps one in his car. He is still a voracious speed-reader with a near-photographic memory.

Phil got hooked on the idea of the underwater world after reading a 1949 *National Geographic* article about a California craze called "goggle fishing." Determined to build his own mask, he fashioned goggles using foam and pieces of plastic window from an old convertible, gluing it all together with what he says was called "Gasket Goo." Although this effort was totally unsuccessful, for a brief second he could see underwater. "That was enough to keep me at it. The idea of going underwater quickly became an obsession. I knew I desperately wanted to do this!"

In 1954, at the tender age of 13, Phil Nuytten became the seventh, and youngest, diver in the Vancouver Skin Divers Club. After Grade 10, he dropped out of high school and started a dive shop in Vancouver's Kitsilano district. Phil wasn't quite 16—too young to get a business licence—so his mother applied for one for him in her name. Although it was stocked with only a meagre supply of scuba equipment, in 1957 Vancouver Diver's Supply became the first dive shop in Western Canada.

Ironically, the *National Geographic* magazine that so captivated a young Phil Nuytten would feature him as a grown-up diver on its July 1983 cover. Diving in Arctic ice in a *WASP* atmospheric diving suit, working with Dr. Joe MacInnis who led the expedition, Phil was the first to reach the wreck of the *Breadalbane*, one of the British merchant ships sent out to find the ill-fated 1845 Franklin Expedition, which had disappeared while searching for the fabled Northwest Passage. The *Breadalbane* had been crushed by ice and sank in 1853, thus earning the distinction of becoming the northernmost-known shipwreck at the time, at 74°41' North. The icy waters had kept it remarkably well preserved. More recently, Phil Nuytten was featured as one of "Canada's greatest explorers" in the January/February 2020 issue of *Canadian Geographic* magazine.

Besides running his own diving shop, relocated a year later at Victoria Drive and 33rd Street in Vancouver, Phil also took on making wetsuits after securing some patterns from an American friend. Phil's girlfriend Mary, who would eventually become his wife, was kept busy making drysuits at his Kitsilano shop. "We called them 'polar bear suits' because they were all white, made out of the white sheet rubber normally used in hospitals." Soon, Ray DeProy, who had two Vancouver dive stores by then, hired young Phil to make wetsuits for him.

One day, the popular professional heavyweight wrestler of the day, Don Leo Jonathan, came in to Circle Diving, a dive shop where Phil often helped out making wetsuits and repairing scuba regulators. "I recognized him immediately from television and shook his hand. He had learned to dive in California and had a very thin California-style wetsuit; he asked me if I could make him a thicker ⅜-inch one for our cold northern waters." Upon completion, Don Leo was very pleased with the suit and inquired about local diving. "It so happened that I was going to Horseshoe Bay to get a couple of octopuses for the Vancouver Public Aquarium," Phil recalls, "so I came up with the idea of taking an underwater photo of a famous wrestler grappling with an octopus. The picture made the local newspaper and that made Don Leo very happy. The same picture also won a prize in the Northwest underwater photo contest and that made me happy."

By the time he was 17, Phil often had back orders for more than thirty wet suits, mostly for the Canadian Navy dive team at Esquimalt, BC. Realizing that running a business had taken over his life, he sold the dive shop and went back to high school at Burnaby Central, earning money by freelance diving on the side. Aside from small salvage jobs, much of the work came from the fishing fleet when net lines tangled around a boat's propeller. Small jobs turned to bigger and more complex underwater tasks, and Phil became known around the commercial diving circle as a dive tender, as well as a hardworking diver in his own right.

After a second stint in school, Phil returned to the world of diving full time, but the companies that used seasoned helmet divers had nothing but scorn for scuba divers and wouldn't hire him. So Phil bought a second-hand US Navy MK5 brass helmet and taught himself how to use it. He wangled his way into Local 2404, the Pile Drivers, Divers, Bridge, Dock and Wharf Builders Union of BC and went to work as a commercial construction diver.

Determined to succeed, Phil did jobs for a number of commercial diving companies, including Universal Diving, down the street from his old dive shop. He also put in seven-day work weeks at BC's logging camps and pulp mills. But the world of commercial diving was changing, particularly as diving jobs moved to ever-increasing depths.

Phil recalls an early commercial diving job at the Bennett Dam on the Peace River in northern BC. Construction began in 1961. When the dam was finished in 1968, it was the largest hydroelectric dam on the Peace River and one of the biggest earth-filled dams in the world. Getting the dam completed on time and on budget involved more than twenty unions signing 10-year contracts promising no lockouts or strike action. There were only a handful of divers and the work was deep at 280 ft (85 m); that meant bottom times of much less than an hour, punctuated with decompression, but there was no decompression chamber available in that area. Phil and Don Leo had already split the cost of a chamber in order to have one for Phil's dive shop, so they simply hauled it to the Bennet Dam job-site.

The area where Phil was diving hadn't been logged, so there were tall trees standing underwater, loaded with silt from the runoff, as the site was flooded. On his first dive, Phil landed in a tree, with his hose tangled in the branches and dirt everywhere as he struggled to get free. A later job at the dam was figuring out why an outflow valve wouldn't seat properly. "We fixed that, not realizing we were creating an instant reputation for ourselves. No matter what you were doing on those projects, it was tough, tough work."

As commercial jobs reached deeper and deeper, some diving companies used a mix of oxygen and helium to avoid the effects of narcosis caused by breathing pressurized air at depths greater than 100 ft (30 m). Other commercial diving jobs used saturation, or "sat," techniques whereby divers could work at deeper underwater tasks and then transfer to a decompression chamber until the nitrogen in their bloodstream returned to normal levels. Regardless of the technique used, safety rules were loose. With costs for this specialized diving incredibly high, the pressure was intense for divers to complete their work as fast as possible. Phil Nuytten developed a reputation as a maverick, diving deeper, staying underwater longer, trying new combinations of gases, diving in Arctic waters. He considered it all part of getting the job done.

By 1966, at the age of 25, Phil had saved enough money from non-stop commercial diving to start his own underwater construction firm. Subsequent years of commercial diving would include plenty of ordinary hard-work jobs, but many records were also established. Phil Nuytten would gain oilfield experience with oxy/helium diving, take part in the first saturation dive under polar ice and the first mixed-gas dives under polar ice. In the process, he helped pioneer new decompression tables for Arctic diving and deep diving, as well as gaining experience in ultra-cold Arctic waters. He would also work and dive with legendary fellow Canadian Dr. Joe MacInnis, who was responsible for the construction of two under-ice habitats in Canada, among many other accomplishments.

Gaining all this knowledge was not without risk. Phil came close to drowning on several occasions—the first time when he was only 18 and a hot-shot spear fisherman taking part in a breath-holding competition in Horseshoe Bay. Phil also ended up with a broken nose when his novice dive tender accidentally dropped a heavy Mark V helmet onto his face. And he suffered vestibular bends during a 600-ft (180-m) dive in 1968, leaving him partially deaf. Decades later, his hearing was further damaged by a faulty radio during an exploration of the immortalized sunken ship *Edmund Fitzgerald* in Lake Superior. Deep, dark and dangerous indeed.

But the hazard of the commercial diving business that hit Phil hardest came when his diving partner Norm MacDonald died in a decompression chamber fire. "That's a terrible way to go," he says quietly. That accident transformed Phil from an innovator to an inventor. Having racked up almost

10,000 underwater hours, Phil decided to give up his focus on commercial diving and consider a different kind of underwater work. Not surprisingly, one of his first projects was related to the cause of MacDonald's death. Phil came up with a device to expel the pure, explosive oxygen that divers breathed and exhaled while decompressing in a chamber, instead of letting it build up inside the chamber. It was just the first of many inventions, some of which, like the *Newtsuit* and the *DeepWorker*, would become known and used around the world.

CHAPTER 3

HYCO: "Where we all started"

1964–early 1970s

———

Don Sorte, Al Trice and Mack Thomson were an unlikely mix. Two of the three, Don and Mack, came from Seattle. Al Trice was the only BC native and is the only one still alive. Now in his early 90s, Al no longer sports his characteristic blond hair or his ever-present pipe, but he is still pragmatic, positive and a sharp thinker. And until early 2020 he went to his International Submarine Engineering office five days a week. It's clear that work is still his life.

Back in the mid-'60s, even before they found a first workshop, the trio needed a name for their venture and ultimately settled on Al's idea of International Hydrodynamics. It was a mouthful. Employees soon shortened it to HYCO. Few would have predicted that this tiny hard-scrabble company would generate global business, thus living up to its "international" name, or that it would spawn a hundred subsea companies with multi-million-dollar revenues. International Hydrodynamics would also serve as the training for a generation of entrepreneurs, scientists, engineers, inventors, designers and technicians. Employees agree that "HYCO was where we all started."

———

All three partners were divers, but Al and Don had also worked together as hardhat commercial divers. They would continue doing so, earning money so they could develop and build their miniature submarine, which would eventually be named *Pisces I.* "We were partners from 1964 on, and like any group we had some arguments," Al notes. "Mack was more interested in inventing things; Don in making money. I was all about practical design engineering."

Al Trice, Mack Thomson and Don Sorte perch on the skeletal structure of *Pisces I* at Vancouver Iron Works, a company on the south side of False Creek that manufactured propane tanks. Personal collection of Al Trice and Pete Edgar

Early HYCO employees have their own memories and opinions of the three diverse partners. All agree that Don Sorte was the front man of the company; and, for a brief time, he was officially president. He was also "the promo guy," "a bullshit artist," "a con man," and "a wheeler dealer." When he came across the Canadian border, he was 25, handsome, bald but with a full beard, married with two children, and continually short of money. One employee says, "Don was a charismatic character, and that magnetism carried a lot of HYCO in the early days." Another recalls that he was an excellent photographer, with an arsenal of Hasselblad cameras he used to good effect.

Pete "Peetie" Edgar, the first HYCO employee, recalls Don Sorte's exuberance. "Life to him was a con job. He'd look at a problem for five minutes and then move on to the next thing. One typical story: Al had put Dubbin wax on his wet work boots and was drying them out slowly in the oven, set at 100 degrees. Don decided to do the same thing, but he was impatient so cranked the oven up to 400. Of course, that didn't work so well." Don loved dogs and had several, including several poodles, all expediently named Tiger.

Fellow diver Terry Knight recalls, "I could walk up to Don any time and say, 'I need 35 bucks.' Fishing it out of his wallet he'd say, 'Don't forget to pay me back.' Then he'd write a little note in his black book."

All HYCO employees agree that Al Trice was the guy with the water smarts. He knew about building and running boats, diving, and all kinds of marine things. "He's the one that really got the first *Pisces* working," Mike Macdonald states. Pete Edgar agrees. "If there was a problem, you could count on Al to come up with a practical solution. Mack would come up with a solution, too, but it would be a much more complex one." Al's constant companion at HYCO and on all dive jobs was his poodle Charlie, who knew everyone. Al taught him to climb wharf ladders and once even took him for a dive in *Pisces*.

Mack Thomson was the inventor, the mad scientist. What he lacked in engineering background, he made up for in determination. Employees and partners alike remember he was always up in his office with the door closed, totally absorbed in his work, especially if the task was mechanical design. Not surprisingly, he was considered the quietest of the three founders. Terry Knight adds, "Mack was very, very smart. And he looked like a very, very smart man, with glasses thick as Coke bottles. He wasn't a big man, yet he was almost as big a character as Don but in different ways. He liked to play with the ladies, as well. And he was always tinkering with projects, like automating the draperies in his house. He really had a terrific imagination and was the guy who did the first test dive in *Pisces*." Mack came up with the name *Pisces* for the first submersible because that was his Zodiac sign. And the idea of Zodiac names for HYCO designs stuck.

Al Robinson recalls his early impression of HYCO's mighty threesome: "I had never met people like Don and Mack and Al. I worked directly under Mack as chief hydraulics technician and he was my immediate god. Once we were discussing a project and I commented, 'Jesus, Mack, isn't that impossible?' He gave me the nastiest look and stated, 'That's not a word we use around here.'"

The combined, diverse talents of Sorte, Trice and Thomson launched International Hydrodynamics. Initially, HYCO had just these three key people:

Mack was usually the pilot for *Pisces*, Don concentrated on diving and never went in the subs, and Al Trice looked after the surface business.

"It took a bunch of crazy guys like Al Trice and Mack Thomson and Don Sorte—and their wives—to change submersible history," HYCO engineer John Witney states. "Those guys would do all sorts of peculiar things just to build their inventions. There were stories of a wife coming home one night to make dinner, turning on her blender, and discovering it wouldn't work. That's because her husband had borrowed the motor to put into a system on the submarine. For Mack Thomson's wife, a conjugal visit in Seattle concluded with the sacrifice of her sewing machine motor, which would drive the scrubbers controlling the critical carbon dioxide level in the sub's cabin."

Once *Pisces I* was finally complete, the insider joke was that the real cost of the submersible was $300,000 and three wives.

Then, as now, Al Trice is a detail man. Fortunately, his memories of the early HYCO years are comprehensive, exact and often salted with humour. "During the years when *Pisces I* was under construction the project ate money, so Don and I started commercial diving like crazy to pay the bills," Al recalls. "Our company was Northwest Diving, and that's what we did for the next five years just to stay afloat."

Many of those diving jobs were unconventional, to say the least. HYCO diver Terry Knight recalls that "back when Don and Al were diving together, the rules were nowhere near like they are today. It was more like the Wild West. Of course, there were already rules that divers should follow about depths and that kind of stuff. But those guys could have written those books because they were pushing themselves past any limits that anybody had tried."

When they weren't diving for dollars, Al and Don moved the *Pisces* project forward, learning on the go. "Mack wasn't an engineer himself, but he knew a couple of Boeing engineers. That's how we connected with Warren Joslyn, who later drew up the engineering design for *Pisces II*," Al states. "We were so naïve at first that we didn't even realize that the sub's design had to start with a sphere in order to survive the deep depth we wanted.

"Mack was a true inventor with absolutely no concept whatsoever of time or money. If we'd left him at it, he'd still be building *Pisces I*. Luckily, we connected with Alex Longo of Vancouver Iron Works, on the south side of False Creek, who assured me that they knew how to build propane tanks so could

Al Trice stands in the partially finished main hull for *Pisces 1*. Personal collection of Al Trice

certainly build a small submarine sphere. They were great to work with and quite keen about the whole project. Also, they let us pay in instalments because we didn't have much money. Mack commuted home to Seattle on the weekends, often talking with Boeing engineer Warren Joslyn, who advised him on the size and thickness required for the main sphere. Mack spent the rest of the time working with the engineers at Vancouver Iron, making the drawings for our submarine—all 44 of them!"

Soon, Don and Al realized that Mack might never finish *Pisces*, so before he got too far along, the two decided they'd better test the main circular sphere and the smaller tail sphere, connected together by a pipe frame. The best way to do this pressure test seemed to be lowering it to the bottom of Jervis Inlet— at 730 m (2,400 ft), it was the deepest fjord on the BC coast. They rounded up their buddies from pile-driving outfits, some volunteers, a crane and a tugboat. Radio expert Jim Spilsbury brought his power boat *Blithe Spirit* to provide accommodation, and the whole group trundled off to Jervis Inlet.

"We had no strain gauge system," Al recalls, "so Mack rigged up a couple of lathes so that they overlapped one another, across the hull. They were taped, but with a little pressure they could move this way and that. He put some cotton batten around the viewport, and we had a big pile-driving hammer for a weight on the bottom of the sphere. That was our extensive test system.

"We rigged the structure to the crane, lowered the thing to the bottom, and it came back up in one piece! Warren, our Boeing engineer, had predicted probable compression at that depth and, by golly, it had compressed just that much."

Elated, the group trooped back to Vancouver, and Mack began the next stage of assembly. HYCO found a workspace on Powell Street at the back of an old mushroom canning factory owned by Harry McGee. HYCO traded rent for occasional stints packing boxes of tinned mushrooms into semis. After several months of mushrooms, the company relocated to a machine shop in the False Creek area that had a bigger workspace—and a better smell.

HYCO's next big job was to construct the fibreglass fairing, which is essentially the sub's outer housing to make it more streamlined in the water. For later submersibles, the company used a plug-and-mould process, but the first time Mack simply fabricated a chicken-wire frame around the main and tail hulls. Then he applied a layer of plaster over that and sanded it smooth. Next, he used resin to glue on layers of fibreglass. Once the hatch openings were cut out, the chicken wire and plaster were removed. It was a labour-intensive job, so in early 1966, HYCO hired their first employee, Pete "Peetie" Edgar.

"We were all young and up for anything," Pete recalls. "But there wasn't much of a paycheque—when there was one. At the time, I was working in The Cave nightclub in Vancouver, running the spotlight. Don was seeing the girl who took money at the door, and I was dating the cigarette girl. HYCO needed somebody to do some wiring, among other jobs, and Don figured I could do that since I had one year as an electrical apprentice under my belt from when I worked up at the big Ocean Falls pulp mill. Even so, wiring a six-and-a-half-foot sphere with no flat walls was definitely a new challenge."

Besides sanding fibreglass and doing some of the wiring, Peetie was a third hand when required and babysat the phone when the guys were out chasing money. Because he was inexperienced, he spent some time correcting mistakes that were the result of not appreciating the harshness of saltwater. "I did learn some helpful tips, like how to smuggle special cable connectors and cans of WD-40 that were only available in the States back across the border. I stayed with HYCO until 1968. By then I'd gotten married and figured I should settle down."

Al Trice describes the submersible's second baptism: "The next time we tested *Pisces*, the sub was pretty well together, including the fibreglass fairing. Don and I had been badgering Mack about another water test, but his answer was always, 'Just another week, another month.' Finally, one day Don and I showed up, loaded the sub on a truck and took off, with Mack running alongside."

They hauled it over to Vancouver Pile Driving's dock in North Vancouver and put it in the water. Whoops. It floated tail down. Looking at it hanging down in the water, Al figured that wasn't such a big problem. It just required some buoyancy on the tail. First, the trio tied a 45-gal (170-l) drum to the tail and sank the sub again, but not too deep. "That's how we learned that a 45-gallon drum can handle internal pressure, but not external pressure when it's empty. It collapsed, and we were tail down again."

Then Al remembered about Grimsby trawl floats from his salmon-fishing days. "I knew they were rated for something like 3,000 feet, so we bought a bunch of them and attached them all around the back end. The main sphere was fairly buoyant, particularly since there was no manipulator arm, no motors that worked, and no sonar, but the good news was that the floats worked and the tail sphere was now buoyant, too."

At some point between the building of *Pisces* and getting it functioning, Al realized something was missing—a launch and recovery system. "Dick Meyers ran Vancouver Pile Driving—and he was long suffering. Of course, we did a hell of a lot of diving for the company, so he agreed to let us construct a gantry on the end

Mack Thompson and *Pisces I* on the gantry at the end of Vancouver Pile Driving's dock in North Vancouver. The steel structure became HYCO's workshop, and the I-beam gantry provided the means to both launch and recover *Pisces*. Personal collection of Al Trice

of his dock." One of the engineers designed a steel structure with an I-beam that went out over the water, which would allow them to pick the submersible up and launch it. The whole affair was covered in and became HYCO's workshop, complete with lights, heaters, battery chargers, air compressor and work bench. "Turns out we were on the end of that dock for eight months."

Initially, Mack lived with Al for a while, then he bought a house in North Vancouver and put a machine shop in the basement. "When we needed a level gauge or some other gizmo, he'd work all night and the next morning come back to the shop with his treasure. Remembering the lessons that I'd learned the hard way back when I was working in the shipyard, I'd take it, look it over, and drop it. Crash! The first few times that Mack bent over to pick up the pieces, he practically had tears in his eyes. But I'd tell him, 'Does it still work? If it survived that, we can use it, because out on the ocean it's going to take a beating.' It took a while, but he finally got the message."

HYCO's procedure for getting *Pisces* functioning was to modify whatever did not work. Then they'd launch the sub, trial the new system, recover the sub and start over again. "The electrical system was always giving us fits," Al says. "Our steep learning curve with batteries was a good example of how we had to trial and rework equipment. Early on, I talked to several battery manufacturers, but none would make a custom one-off battery for *Pisces I*. Then Joe Hartle, another diver, suggested getting in touch with Ernie Davidson of Davidson Batteries in Vancouver. Ernie was a one-man operation making lead acid batteries for retailers and for fishing boats. He listened to me and said, yes, he could make our battery. The result was a battery made up of 60 cells, which resulted in 120 direct current volts. The cells were housed in a waterproof box filled with transformer oil, which floated on the battery acid. Since the box was full of liquid, it was immune to sea pressure.

"*Pisces I* had two battery boxes, one on top of the other, both made from aluminum. One time we discovered a ground fault [short circuit] in the top box. But the batteries were already charging ... so we decided to drain the oil, remove the box lid, and fix the ground fault. We quickly learned that was the wrong decision! That's because when batteries are charging they generate hydrogen, which vents off through the oil. Once there was no oil, the ground fault arced to the aluminum and ignited the hydrogen, which exploded and blew the box *and* the top of *Pisces* to hell." That's when Al decided no more aluminum—from now on, the boxes would be non-conductive fibreglass. It took a lot of work and testing to make a functioning box, but finally that problem got solved. Through it all, Ernie Davidson was a key player for HYCO. He went on to build the lead acid batteries for all the subsequent *Pisces*.

Finally, the day came when HYCO had a working submersible! Of course, a few items were still missing—like an underwater telephone, a sonar, a directional hydrophone, a pinger, cameras with a pan-and-tilt mechanism, and manipulators. But over time, HYCO hired engineering specialists to develop much of this equipment. For now, Don, Mack and Al were glad that their sub worked! *Pisces I* was only 16 ft long with a beam of 11 ft (4.8 × 3.3 m). The pressure hull was just 6.5 ft (2 m) in diameter, just enough space for a pilot and observer. The depth rating was 1,500 ft (460 m), although the submersible had been to 2,400 ft (730 m).

In the mid-'60s, there wasn't much subsea equipment available for HYCO's small scale of vehicles, so the team had to improvise. Al Robinson did all the early hydraulics for *Pisces*, but there was no sonar for a small submersible and no manipulators. The first sonar system *Pisces* used was a fishing sounder from the Seattle company WESMAR. HYCO knew manipulators were also necessary, but they just didn't exist yet. So the company contacted Jack Russell, who ran Progressive Hydraulics and got him to fabricate the sub's first hydraulic manipulator. After that, Al Robinson built them for HYCO, adapting the design as necessary.

During this time, HYCO was not oblivious to the fact that they owed lots of money. "By the time we got *Pisces* operating, we were in debt $60,000." Money was extremely tight, so the threesome decided they needed a friendly bank manager. Al says, "We went through an entire list of bank managers and came up with only two that might put up with us. One of them was Bob Pope, with the Bank of Montreal. He listened carefully to all that we had to say and finally said, 'You know, I'm really intrigued, so I think I'll fly with you guys, but your limit is $2,500.' So we just went ahead in $2,500 clumps."

During this period, the learning curve was incredibly steep. The HYCO team was trail-blazing. There was nobody around with comparable experience and no book had that kind of information. "We had to create everything from scratch with no real understanding that it would be reliable or even that it would work in the ocean. Everybody we hired had to learn the concepts of underwater engineering right from the start," Al says. "What comforted me was the thought that if it was easy, everybody would be doing it."

Back in the mid-'60s, Vancouver newspapers had marine reporters who covered the waterfront. From the beginning, they were always writing articles about these crazy guys building a submarine. HYCO fostered this coverage. Once *Pisces* was operational, any interested party with money was taken for a dive. Those newspaper articles provided terrific free advertising for the cash-strapped company.

While *Pisces* was still under construction, a couple of geologists from Calgary came to HYCO. They were looking to use the submersible for a geological survey in Hudson Bay for Richfield Oil in the summer of 1965. Al explained that *Pisces* wouldn't be ready by then, but took the time to discuss the kind of boat they would need for their project—a high-speed live-aboard that could carry maybe five guys doing a survey for several months. The problem was that such a boat didn't exist, so he introduced them to North Vancouver boatbuilder Loren Brown. They signed a deal and construction immediately got underway. As a result of his input, Richfield Oil asked Al Trice to run the entire survey operation as well as scuba dive in the shallow areas of Hudson Bay in order to map the area and retrieve samples. HYCO needed the money, so Al took the job.

The *Hudson Explorer* easily transitioned from conducting sonar survey work in Hudson Bay to serving as towboat for HYCO's *Pisces* submersibles. Personal collection of Al Trice

The challenges soon became apparent. First of all, there were aerial photos but no marine charts for most of Hudson Bay. When Al flew up to show Richfield Oil where to set up fuel dumps, he realized the vast distances involved. He called Loren Brown and told him to add an additional fuel tank. Then another problem arose. Brown let Al know that he was going bankrupt and worried that the boat might be seized by creditors. "So one night he tied the boat up and we did a midnight flip. We loaded the boat on a truck, along with the tools we needed, and hid it, completing it elsewhere!"

Once completed, the boat had a freezer for meat and room for four men—Al Trice, Jim Kail (a second diver Al had recruited), geologist Bill Mackesy, and sonar operator Neil Wilson (who joined up after a night of too many martinis). In the end, the four of them lived and worked on the boat for four months. They looked at little else apart from each other and lots of ice, all the while sonar mapping and diving in some very cold water.

When the survey finished, the four packed up the gear and put the boat up on dry land for the next season. But then Richfield Oil decided against any more surveying in Hudson Bay, so suddenly the boat was for sale. Al bid on it and won. He loaded it on a flatcar in Churchill and brought it home. "The *Hudson Explorer* became our towboat for *Pisces*," Al Trice explained. "Without it, we would have been screwed."

Finally, the launch of *Pisces I* and a first job

Pisces I was launched in 1966 at Vancouver Pile Driving with no official fanfare. In fact, only Al Trice, Don Sorte, Mack Thomson and a photographer were in attendance. The low-key event was simply the next step for HYCO: getting the submersible in the water. It would take eight more months to get *Pisces* completely functional.

Pisces got a first contract even before the sub was fully outfitted. In 1967 Bill High, who worked for the US Fish and Wildlife Service, and the University of Washington agreed to pay $2,500 to acquaint biologists with observing marine life in situ. Suddenly, HYCO had to get organized. They rented a beat-up, 90-ft (27-m) wooden scow from their buddies at McKenzie Barge and Derrick and put their gantry on it. Then they added an electric hoist, financing it on the never-never plan—a dollar down and a dollar someday. It helped that many maritime companies were intrigued as hell about this crazy submersible taking shape. Next, a friend surrendered the engine from his Mercedes to run a generator that would charge the sub's batteries and run the hoist.

The official launch of *Pisces I* was a quiet celebration; Al, Mack and Don knew there was still plenty to do for the submersible to be functional. Personal collection of Pete Edgar and Al Trice

"Finally, we had the whole thing ready—the barge, the sub and a boat to tow it," Al recalls. "We decided we'd just charge Bill High $2,500 to tow *Pisces* down to Seattle and use the trip as publicity. Peetie Edgar went along on this job but took a day off to get married!" The entourage stopped in a couple of places on the way so that biologists could dive on *Pisces*. "They were absolutely flabbergasted! In fact, we had a hell of a time getting them out of the submersible. They'd never glimpsed all these critters alive in the sea before. All they'd ever seen was what had gotten dragged up dead in plankton nets. One scientist was so excited he kept talking into his light meter instead of the microphone of his tape recorder."

HYCO won another early paying job with US electrical engineering giant Westinghouse. Like Rockwell, Lockheed and Phillips, they were considering a move from aerospace work, which was winding down, to *hydro*space. Having built a bigger submersible of their own, Westinghouse wanted to acquaint their people with the operation of a manned submersible. So they hired HYCO to take employees down in *Pisces* in Saanich Inlet on Vancouver Island. "Everybody loved it!"

While eager for early work, HYCO was still building its own workforce. Mike Macdonald was a good example of how informal their hiring was.

After getting a job in the physics department of the newly minted Simon Fraser University, Mike had also enrolled in a scuba-diving class. That's where he heard a guy say, "If you want to send someone down to 600 feet, there's 600 instruments we need and only 20 of them exist." Intrigued, Mike started reading up and discovered there was a company in North Vancouver making a submersible. He visited HYCO and found Al Trice kneeling on the floor lofting the new design for *Pisces II*. "I started hanging around a lot and spent time in the sub, tracing wires just for the hell of it, and taking out the garbage and whatever else I

Pilot Mike Macdonald waits in the open hatch of *Pisces I*. Personal collection of Al Trice

could do. Eventually, they said to me, 'We're going to California next week with the sub and it looks like you've kinda figured out how things work. Why don't you come with us?' So I quit Simon Fraser in the fall of '68 and joined HYCO."

Spending two months in Santa Barbara demonstrating *Pisces'* capabilities to oil companies was a dream job. Mike recalls: "I mean, hell, could a young guy ask for anything better? We probably took hundreds of people underwater and they all loved it. Then just as we were finishing, there was a big oil spill in California that shut down the whole oil and gas industry there. So we came back to Vancouver just as *Pisces II* was completed. In those days, there was no such thing as pilot training. I got started one day when we were at Vancouver Pile Driving's big dock, doing some tests and putting subs in the water. They said, 'Mack ain't here, so I guess you're the pilot.' That's how I started."

Working the torpedo range and paying the bills

On one Seattle business trip, Al and Don met an engineer in the Applied Physics Lab at the University of Washington who had designed all the arrays for the Maritime Experimental Test Range. That was the joint venture between the American and Canadian navies located at Nanoose Bay on the east coast of Vancouver Island. That meeting was critical for HYCO, giving them the opportunity to work with the US Navy. Beginning in 1967 and for the next several years, torpedo recovery would provide steady work, allowing the company to pay off the $100,000 cost of building *Pisces I*.

The Americans had just developed the Mark 46 air-dropped torpedo that became the backbone of their Navy's lightweight anti-submarine torpedo arsenal. These new torpedoes only weighed 500 lbs (227 kg) and were 12 ¾ in (32 cm) in diameter. Normally, they carried a high-explosive charge, but the ones in the test range had a different head with a pinger inside for tracking. After being launched they'd run for 20 minutes, burn up their fuel, and then rise to the surface. But if they stopped running while still heavy with fuel, they'd sink. At the time, they were sinking left, right and centre. And since these torpedoes were *very* expensive, the Navy needed to recover the duds so they could analyze the problem.

A Seattle outfit had been doing some torpedo retrieval but only had a ponderous grapple system along with an underwater camera. And it could take a couple of days just to recover a single torpedo, so the US Navy wanted to try *Pisces*. In order to do that, HYCO had to develop a claw that could pick up the torpedoes. The company did so, practically overnight. When the first one didn't work, they trashed it and built another.

The US Navy also required that *Pisces* pass several tests. The real action evaluation was conducted on the torpedo range in Nanoose so the US Navy could see if *Pisces* could find one of their torpedoes. Al chuckles, "They gave us a pinger locater, and then were going to give us an expensive underwater telephone system, but it wouldn't even fit through the hatch of our submersible! Instead, we got ahold of an underwater telephone from California. With that and the pinger receiver the Navy loaned us, we were able to find and pick up these torpedoes with the sub."

Once the Navy decided *Pisces* could do the job, they required a final test dive to 1,800 ft (550 m), since they were also running torpedoes in Jervis Inlet, which was deeper than the Nanoose torpedo range. The story of that very first deepwater test dive for the US Navy, conducted up in Jervis Inlet, quickly became legendary.

Mack Thomson was piloting the sub by himself and had a tape recorder running to document his observations; Warren Joslyn's strain-gauge wiring was strung all over the co-pilot area. Once the weights were added, the sub started to descend. Every 200 ft (60 m) or so, Pete Edgar was on the radio to Mack. "Everything OK?" Although the response was garbled, Mack would confirm, "Yeah, I'm at such and such a depth and everything's all right." So far so good. But underwater communication was only marginally effective and heavily dependent on HYCO's *Hudson Explorer* sitting directly above the sub. The increasing difficulty in understanding Mack made the surface crew realize that *Pisces* had drifted out from under them and was moving closer to the nearby US Navy vessel that was launching the test torpedoes.

What worried Al Trice was that the vessel was still firing live torpedoes with *Pisces* in the water. Normally, homing torpedoes were air-dropped from a ship and would head out to a test target ship, which had an electronic target lowered overboard. The torpedoes wouldn't actually hit this target because there was a turnaway circuit to prevent that. But the reality at this moment was that they were anti-submarine torpedoes—and *Pisces* was a submarine.

That scared the hell out of Al, so he made a quick radio call and told the Navy to stop firing the torpedoes. Reluctantly they agreed, but there seemed to be some delay in the message getting through to the firing crews as the men on the *Hudson Explorer* spotted the ship launching two more in quick succession. With no time to call off the dive, Al immediately contacted Mack, advised him what was happening and told him to shut off all the electronics he could aboard *Pisces*.

Mack heard enough of the garbled message to realize his predicament. He shut down everything possible with the exception of the tape recorder. He even covered his wristwatch to muffle its ticking. Then he waited in the silent sub. His tape recorder captured the whine of the torpedo propellers cutting through the water, seeking a target. Fortunately, it was not *Pisces*. Al says, "To

Mack Thomson examines the collapsed ballast sphere from the tail section of *Pisces I*. Personal collection of Pete Edgar

this day, we don't know how close those torpedoes came."

Eventually the danger passed and Mack continued the test dive. However, as the sub drifted farther from the *Explorer* all communications ceased. On the surface, Peetie kept saying, "Hello Mack. Do you read me? Do you read me? Mack, are you there?" Nothing. Topside, the crew had no idea what was going on down below or even what could be done about it if they did know.

Shortly after *Pisces* passed 1,500 ft (455 m) in depth, Mack was startled by a loud bang. At the time, he didn't know that a ballast sphere had imploded. But he did realize that *Pisces* was sinking.

He immediately blew the ballast tanks, but it wasn't enough to stop the descent. As the depth gauge registered the sub's plunge, Mack began struggling to release the 185-kg (400-lb) dropweight, which was attached through the centre of the submersible by single-threaded shaft. But the increasing water pressure compressed the hull so that it was now binding on the shaft. On the tape, Mack calls out each new depth rating, then puffs and pants and grunts as he struggles to loosen that dropweight. Finally, it released. "Thank Christ. It's gone," the tape records.

Then there was silence as Mack waited and the sub's decent slowed. Finally, a further 30 m (100 ft) deeper, *Pisces* came to a standstill. Ironically, the depth was 600 m (1,970 ft), slightly deeper than the Navy required. Mack continued staring at the depth gauge. Then slowly, very slowly, he realized the sub was starting to rise. On the tape you can hear, "Oh, I think I'm coming up. Yes, I'm coming up, thank Christ."

Despite the near miss with a test torpedo and an imploded ballast sphere, both Mack Thomson and *Pisces I* survived and passed the us Navy depth test for torpedo recovery work. After recovering the submersible, the crew celebrates with champagne. Left to right: Don Sorte, Mack Thomson, Peetie Edgar and Al Trice. Personal collection of Al Trice

At the surface, time dragged and darkness had fallen with no word from *Pisces*. The Navy vessel had departed; it was raining and totally black. Al Trice directed everyone back to the barge and then fired up the generator in order to get lights turned on, knowing that *Pisces* would need a reference point. "I also knew the tides, and there was a slight current running out of the inlet, so I hoped that was where we might see our sub," Al recalls. "Then we just waited.

"All of a sudden, *Pisces'* big lights illuminated the water, and Mack was back on the surface! We hooked onto the sub and towed it back to the barge. When we got there, Mack was full of things to say, understandably. But in the end, everybody agreed: So what if the ballast sphere collapsed? We got the sub up, Mack's OK, and we passed the test. It's a success. Pour the champagne!"

When HYCO first started recovering torpedoes for the us Navy in August 1967, there were many failures, Al recalls. "I was upset about this until I realized that the operational equipment the Navy used for other work on the range had the same if not a greater rate of failure. Fortunately, we learned pretty quick. And amazingly, they paid us $2,500 American for one torpedo. That saved our bacon because we were pretty heavily in debt."

Torpedo recovery was always a challenge. The Navy would fire their torpedoes during the day, with all of them meant to come back to the surface after being fired. HYCO's job was retrieval of the ones that didn't. "That was all night work," Mike Macdonald recalls. "We'd be working in North Vancouver and get a call—there weren't cell phones back then—saying 'torpedo's down' so we'd jump in the car and get on the ferry. The secretary would call my wife and say, 'Mike won't be home tonight. He's over picking up torpedoes.'"

The work at Nanoose Bay became HYCO's bread-and-butter contract over a period of three years, first with *Pisces I* in 1967 and then with *Pisces III* in 1969. Although the naval base was in Nanoose Bay, the actual torpedo operation base was on the Winchelsea Islands, a group of islets some 2 mi (3 km) outside the mouth of the bay, with HYCO's barge tied up at a nearby mooring buoy. "Depending on whether you were working with the diver or in the submersible that night, you'd get on board the towing boat or into *Pisces*," Mike says. "There was no way to get into the submersible offshore; it was always too rough. The boat towed *Pisces* backwards to the dive site so the line wouldn't foul the thruster props. Sometimes the tow took two hours, so we'd be sitting in the sub huddled up in a sleeping bag. It was cold, and *Pisces* didn't tow very well, twisting one way and then the other."

The range at Nanoose Bay had a one-mile grid, marked out with what were called arrays—steel spheres with electronics inside that allowed the Navy to track any torpedo that didn't surface and relay the rough coordinates where it fell. "We'd tow out to where the Navy said the torpedo had gone down, dive the sub and start looking for the failed torpedo, but it was hardly ever where they said it was. Fortunately, we had a directional hydrophone on *Pisces* and the torpedoes had pingers, so we'd turn the sub and listen, then go to where we heard the loudest sound. Hopefully we also saw something on the sonar. It was only a fishing sonar because that's all that was available cheaply. That sonar was why I started doing more diving in the sub. I was told 'You understand physics, so you're going to operate it.'"

Pisces was equipped with a manipulator with two removable "jaws"— little clamps for the smaller Mark 46 torpedoes and big clamps for the Mark 48 ones. Once the torpedo was located, the *Hudson Explorer* radioed the Navy's recovery launch. They would come out and either grab the line or put a nose cone on the torpedo and haul it onto their boat. "Then we'd be done unless there was another one to find."

The Mark 46 torpedoes were fairly easy to pick up once the *Pisces* crew found them, but the Mark 48s usually buried themselves in the mud, often with only the back fins visible. "Those were much harder to recover, but that's how we learned a lot about pumps and how to dig holes. Eventually, I guess that led to HYCO getting into the cable burial business."

Al Trice laughs as he recounts recovery tactics. "We had a lot of bad weather on the range, and those Navy launch crews were reluctant to come out in it, which meant we couldn't hand off the torpedo. So I bought a case of rum and every time they came out to get a torpedo in nasty weather, I'd pass them a bottle of rum. That solved that problem pretty quickly!"

Al also recalls, "The Navy used all kinds of acronyms when they'd transmit over the radio. The range officer was 'Polar Tanker,' torpedoes were 'units,' and the range boats all had individual code names. It was a real communication carry-on. So, one time when we were going to contact *Pisces*, Peetie got on the radio and said, 'White chick, white chick, this is fussy hen.' We just about split a gut over that one!"

The Naval torpedo station at Keyport in Washington state had a discretionary budget of $10,000 and gave HYCO sequential $2,500 payments per recovered torpedo. Al still smiles remembering, "In no time at all, we had paid off the $100,000 we owed on the submersible. The bonus was that working with the US Navy was a great education for us."

Torpedo recovery was a coveted contract, but the work see-sawed between the Canadian company and a Seattle-based one. HYCO got its first contract in 1967, then lost the bid in 1968, a blessing in disguise because it allowed the company to concentrate on developing *Pisces II* and *III*. HYCO won the contract again in 1969, but eventually the Seattle company got a remotely operated vehicle (ROV); no longer needing to send a pilot down on each dive, they only charged $1,500 per torpedo. That was the end of torpedo-recovery work for HYCO. However, HYCO's operational development on the Nanoose torpedo range facilitated sales of their submersibles to the UK and significantly accelerated the Royal Navy's torpedo recovery programs.

Pisces I and her crew got their first exposure to working and diving in Arctic waters in 1968. One of the advantages of all *Pisces* craft was that they could fit into a plane for rapid transport. Personal collection of Al Robinson

In 1968, during the Cold War, the Canadian government arranged for *Pisces I* to go to the High Arctic. The Canadian government's Pacific Naval Laboratory at Esquimalt (later called the Defence Research Establishment Pacific) had built listening devices and installed them in all the channels that acted as choke points through which any Russian submarine attempting to come through the Canadian islands into the North Atlantic would have to pass. These listening devices were roughly the size of a desk and had a recorder, a hydrophone, a winch-and-buoy system, and a dropweight. The mission was to dive on them in *Pisces* and take pictures. The HYCO submersible and her crew were flown to Thule, Greenland, aboard a big Hercules aircraft and then transferred onto the icebreaker CCGS *Labrador*. The ship had geologists, biologists and physicists on board, so the work was never boring. Al says, "We learned like crazy working in ice. One of the problems *Pisces* had was that the solenoid valves we used for vent valves would freeze. So before each dive we had to saturate them with anti-freeze. That fixed that problem. And then the o-rings would get real hard, so we tried using Vaseline, and that solved that problem. It was always like that."

Introduction to the oil and gas industry

Aside from military work, securing contracts in the oil and gas industry was a crucial step for any underwater company in the late '60s and '70s. Generally, those contracts paid big money, but they also involved high risk and tight timelines. In 1969 HYCO went to work for Shell Oil, which had an exploratory drilling program with a SEDCO 135, the class of $10-million Canadian-built offshore semisubmersible drilling vessels owned by the South East Drilling Company (hence the acronym SEDCO) and built by Victoria Machinery Depot. At the time, the SEDCO 135-F was drilling in the middle of Hecate Strait, off the coast of northern BC. Mike Macdonald explains: "They had lost a great big bloody piece of pipe—it was the first piece that goes into the hole in the bottom and that the wellheads bolt onto, and they needed it back immediately. HYCO leased an ocean-going tug and loaded everything on the back deck and headed out. When we got there, it was blowing like a hurricane, with winds 75 miles an hour. We circled the SEDCO for two days, unable to do anything." Eventually, the guys put *Pisces* in the water, found the missing piece of pipe, put a choker on it and got it recovered.

Mike continues: "While we were down with *Pisces*, we took a look at their anchor chains. They had piggy-backed 15 ton of anchors, three in a row, to keep the SEDCO 135-F in position. But the weather was so bad that this huge drilling rig was dragging! The week after we got off it, a 100-foot wave went right under that rig. That was our introduction to the oil industry. I mean the *real* oil industry."

A year later, in 1970, HYCO secured a contract with Aquitaine Petroleum, a large international oil and gas company based in Calgary that was a subsidiary of a company from Pau, France. Aquitaine wanted to use *Pisces I* for deeper dives to examine the test wells that had already been drilled up in the Hudson Bay seafloor. They wanted to see if the wells were leaking and to obtain seafloor rock samples. All that work would be happening in difficult, icy conditions and at a water depth of 700 feet, considered "deepwater" in the '60s.

This sketch of the *Hudson Handler* was featured in one of HYCO's company brochures. It shows the recovery ramp for *Pisces* in rough weather. Another important plus was the vessel's segmented construction that facilitated shipping to remote locations. HYCO brochure from Al Trice's personal collection

Having worked with boats in the ice back in 1965 and 1968, Al Trice was well aware that the variable Arctic sea conditions required a different style of boat in order to safely launch and retrieve a sub like *Pisces*. Al Trice met with Serge Rueff, a French petroleum engineer who was the Hudson Bay project manager for Aquitaine, and explained that if they intended to use a normal vessel to launch *Pisces*, chances were they might not manage even one successful launch and recovery in rough weather. Instead, Al suggested that a boat with a dedicated launch and recovery ramp was a much better idea. But it would also have to be built in segments for easy transport.

Aquitaine gave the go-ahead to build such a boat; the only clincher was that it had to be built very quickly to take advantage of the ice-free window. Robert Allan Ltd. in Vancouver did the design for a self-contained, self-powered modular barge that could easily retrieve a *Pisces* submersible. HYCO called it the *Hudson Handler*. Essentially, it would be a much bigger, beefier companion to the *Hudson Explorer*, HYCO's small workhorse that could only tow the *Pisces*.

Al then contacted Ted Purvis of Selkirk, Manitoba, who agreed to build the *Handler* on the tight timeline. Two fabricators in Winnipeg each built half of the hull. When completed, it was 90 ft long (27 m), 40 ft wide (12 m) and weighed 343 tons. Propulsion was supplied by two big second-hand self-propelling units, each with a 671 GM diesel engine. In the end, the boat was fabricated in only 26 days, a remarkable feat that allowed the survey team to get the maximum use of the short Arctic season. The *Handler* and other gear were quickly loaded onto 13 train cars and taken to Churchill for reassembly. Al concludes, "That's often how our work went—some company had a unique job or needed a special service, and our challenge was to deliver it, usually on a tight deadline!"

That first Aquitaine contract was followed by a second one, again in Hudson Bay during the summer of 1971. This job was marked by extremely difficult surface and underwater conditions—rough weather, strong sea currents and cold water, but was completed successfully. Aquitaine's third contract with HYCO was off the US East Coast where oil and gas companies had drilled test wells but had gotten chased off by icebergs, so aimed to return to them the next season. *Pisces*' job was to re-locate these wellheads and inspect them for any damage. Originally, they were marked with pop-up buoys and pingers, but the buoys had gotten entangled and the pinger batteries were dead. The *Pisces* crew used the sub's two manipulators to free the buoys and then added pingers with larger battery packs, successfully completing the contract.

That third Aquitaine contract was the start of pilot training for Tom Roberts under the guidance of Fred Warwick. "Piloting wasn't really tough, but you couldn't just do things spontaneously. You had to plan everything out because you never knew what you were going to run into or get hung up on. So you always had Plan A, B, C, D and F. Learning how to put that together was probably the hardest part. Fortunately, each person you worked with taught you a little different aspect of piloting."

The success of these missions in both the Low and High Arctic led to subsequent HYCO contracts with Total Oil Canada, Petrofina, and Hudson Bay Oil & Gas. The specially designed *Hudson Handler* became an integral part of the services that HYCO delivered on these contracts. Equipped with fuel and support facilities to sustain a crew of 10 for 30 days, its Canadian Steamship Inspection certificate approved it for open ocean work.

Much of *Pisces*' work was global, but two local salvage jobs made BC newspaper headlines. Al Trice planned and managed the challenging field

The *Emerald Straits* was built in 1950, one of the early steel tugs. In 1969 she sank deep in Howe Sound. Three crew members lost their lives; one washed overboard and swam to shore. This photo shows the tug being raised back to the surface. Personal collection of Al Trice

operations for the successful recovery of two tugs—the *Emerald Straits* (1969) and the *Haro Straits* (1972). Both were some of the first steel tugboats built in BC, and both had gone down with loss of life. The BC tugboat union demanded to know why.

The 95-ton, 51-ft (15.5-m) *Emerald Straits* was at 670 ft (200 m) in Howe Sound, an extremely deep salvage depth at the time. The plan for raising the tug was that *Pisces* would cut the two anchor chains on the tug's bow at the windlass and then insert a toggle bar into each of the hawse pipes. Then a sling could be passed from the tug's bow to its stern, providing fore and aft lifelines to the barge anchored above the sunken ship. Since no hydraulic chain-cutter existed, HYCO fabricated one in five days.

After much difficulty, both anchors were removed and one toggle was successfully inserted. But the second toggle jammed in the hawse pipe. Working on the fly, HYCO bolted a 65-lb (30-kg) weight onto *Pisces'* manipulator arm to create a makeshift hammer. Even so, the sub had to manoeuvre back into position after each blow. After 10 hours of pounding, the second toggle was finally inserted, and a successful raising followed.

Three years later, the *Haro Straits* went down in 420 ft of water (130 m) with the loss of all hands. The tug sank in American waters, just off Point Roberts, and the US wanted it removed. Al Trice, divers Scratch McDonald and George Hazelton, *Pisces* pilot Fred Warwick and barge deckhand Tom Roberts, among others, worked on this job. Al Robinson and Merle Wilson were also diving in *Pisces* to check that the rigging for the lift system was lined up and OK. Al Robinson admits, "There were the occasional panic moments, like when we ended up trapped in the sub for about eight hours, tangled in a line at 300 feet. But I knew that Al Trice and George Hazelton were up on the barge—two incredible men I had total confidence in. And sure enough, eventually that problem got solved and we got back to the surface." This salvage work was yet another challenging job that HYCO's resourceful crew and its *Pisces* submersible completed successfully.

A new start and an ending

Al Trice didn't dive in *Pisces I* for several years; usually he was running the surface operation. Then in 1969 he got married for the second time and finally got to dive in *Pisces* for the first time. That's when he realized, "Hey, I'm dry, I'm warm, there's relatively unlimited depth and no decompression, plus the stereo's playing and I've got my lunch. Who wants to go suit diving?! So *Pisces I* was the start of HYCO, but it was also the end of cold-water diving for me. You're damn right!"

CHAPTER 4

HYCO: Chronicle of a subsea legend

Late 1960s–1979

B uilding a submersible, even a small one like *Pisces*, required money. Lots of money. So Don Sorte set up the Northwest Diving Company as the separate funding arm for HYCO. As a result, many early employees joined the company as commercial divers. Others were hired to help in HYCO's shop fabricating the submersibles. Some did both. By 1970, just a few years after *Pisces I* went to work and *Pisces II* and *III* were completed, the firm could boast 55 employees, including divers, a diving school, manufacturing operations that produced more *Pisces*, and two sub crews who travelled with the submersibles for jobs that ranged from burying subsea cables to conducting underwater surveys for the rapidly expanding oil and gas industry. Here's a sampling of early divers and how they came to HYCO:

- **Doug DeProy** got his scuba ticket as a young teenager, trained by Phil Nuytten, and also became a helmet diver. His father, Ray DeProy, had two retail dive shops in Vancouver. Don Sorte lived in the DeProy family's basement suite, so eventually Doug joined Northwest Diving and became heavily involved in the company's Northwest School of Deepsea Diving, which was set up during a slow period of commercial diving.

  "For a couple of years, I was the chief diving instructor at the school. I went down to California in 1970 and certified with the National Association of Underwater Instructors. I taught diving for many years, was president of the Canadian branch of Underwater Instructors for a couple of years, and later my wife and I owned two dive shops and a commercial dive company. Working at the dive school, if a big commercial diving job came up, I'd leave the dive school, do the job, then come back to teach." Doug also did PR work to raise money for HYCO, giving slide shows at men's clubs in order to get investors. He also manned HYCO displays and mockups at the Pacific National Exhibition grounds during its annual summer fair.

"The Northwest School of Deepsea Diving offered a four-month course for eight students, and we did three courses a year. Its success was thanks to the Canadian government because 90 per cent of our students got 100 per cent of their $10,000 tuition paid. Classes were 8- to 10-hour days, five days a week, so there was a lot of

Doug DeProy notes: "In the classroom I was the instructor. In the water I was the safety/standby diver, although I've always wondered how a scuba diver could rescue a heavy gear diver." Personal collection of Doug DeProy

instruction. We taught Mark V hardhat diving and supplied all the heavy gear. In those days you were only considered a real diver if you had a brass helmet. Scuba diving was considered amateur. Part of the cachet of the diving school was that students also got some submersible time."

Students spent approximately 20 per cent of their time in the classroom. There was also a small mechanical room and four underwater work stations, complete with a steel bench, for each dive pair. The maximum enrolment was eight students, which allowed for one diver in the water and one tender, working as an alternating team. Communication was either by radio or more commonly the "line pull" system. "Back then it was still like the dark ages!" Doug adds.

- **Deloye "Scratch" McDonald** was a high-school buddy of Don Sorte's in Seattle. "Don and I were always looking for ways to make a buck besides our regular jobs. I had four children and a wife to support. After I got out of the Army, Don and I became divers, salvaging sunken logs for a sawmill for a while." Deloye's well-known nickname "Scratch" came from the heavy, handknit sweater he always wore under his diving suit for warmth. That sweater featured a large dollar sign that generated his nickname, since "scratch" is American slang for money. "Then Don left to start working with Al Trice and set up Northwest Diving, which was separate from HYCO. Don called and offered me a job, so I moved my family to Canada."

  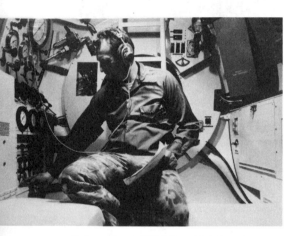

  Deloye "Scratch" McDonald doing a pre-dive check of various *Pisces* components. Personal collection of Deloye McDonald

  Deloye segued from diver to submersible pilot. "When HYCO decided to build *Pisces II*, they needed another submersible pilot. HYCO brought in a few guys, but there was a leak in the battery box and POW!—the batteries blew up. Fortunately, I was working with my head down, because the tank blew off right over my head. That was the last we saw of one trainee! One day there wasn't anyone else to dive with the pilot, so I jumped in the sub. The next day they asked if I wanted to become a pilot. I thought, 'Well, why not?' In those days training was just learning by doing things.

  "It was fascinating work but tough on the family. My wife raised the kids. I'd be sound asleep and the phone would ring saying, 'We need you right away,' and off you'd go. You never knew how long the job would take. It could be hours, or it could be days, but that's what made it interesting!"

- **Terry Knight** was an industrial electrician working at a sawmill and got interested in diving through a friend. "I decided that I wanted to become the next Mike Nelson from *Sea Hunt*, so I looked into the

Northwest School of Diving. They told me I could join, but their eyes really lit up when they heard I was an industrial electrician because they really needed an electrical type in the shop. So I took that job.

"Don Sorte had a theory that you could give someone a title or a salary but not both. I became the West Coast field operations manager, which tells you something about my salary! While I was working there Al Robinson and I became best friends and later we teamed up to start Inuktun in Nanaimo. Of all the escapades in my life, my time at HYCO and with Don are among the very best."

- **Bob Starr** of Seattle originally started out in the diving school. He had already gone to a diving college in Washington state, so commuted back and forth sharing his knowledge. Later on, he became operations manager for HYCO and was Mike Macdonald's co-pilot when they made the first test dive of *Pisces V* to 5,000 ft (1,520 m) off the East Coast. Bob was also Mike's second-in-command at HYCO's Houston office.

Early on, the three owners of International Hydrodynamics also hired technical expertise, beginning with Boeing engineer Warren Joslyn. As business picked up, more mechanical experts came on board, but everyone always pitched in with whatever job needed extra hands. Al Trice states proudly, "A lot of smart guys came to work for us. At one time, we had over a hundred people at HYCO!" The list below is just a sampling of those pilots and engineers and their diverse backgrounds:

- **Pete "Peetie" Edgar** was technically the company's first employee, generally doing what needed to be done, whether that was electrical wiring, sanding fibreglass or smuggling WD-40 across the US–Canadian border.
- **Warren Joslyn**, a Boeing engineer, started as HYCO's first, key consultant. He commuted between Seattle and Vancouver, advising Mack Thomson on the size and thickness of the main sphere for *Pisces I* and producing the technical drawings for *Pisces II*, including the subsequent re-design and refinements in HYCO's *Pisces* fleet.
- **Mike Macdonald** became one of HYCO's most experienced pilots and instructors. In 1973, Mike piloted *Pisces V* during the harrowing 80+ hour

After *Pisces II* was completed in late 1968, Warren Joslyn quit Boeing, working fulltime with HYCO for 10 years. He left when the company wasn't building any more submersibles. Personal collection of Al Trice

rescue effort of *Pisces III* from the bottom of the North Atlantic, as related in the book *No Time On Our Side*, by one of the *Pisces III* crew, Roger Chapman. In 2021, Stephen McGinty authored a new version in *The Dive: The Untold Story of the World's Deepest Submarine Rescue*. Mike Macdonald also became technical manager, project engineer and engineering manager for HYCO and its subsidiary HYCO Subsea Ltd. in Houston before joining ISE.

- **John Witney** helped with engineering on several submersibles and eventually also became a pilot, working in the Gulf of Mexico, the North Sea, eastern Canada and the coast of Africa. "We set a bunch of world records with the deep-diving *Pisces*. I was the first Canadian to go down to 6,500 feet and actually the first Canadian to get back from 6,500 feet!"

- **Les Ashdown** first worked in the shop at HYCO and then became a pilot and instructor on the torpedo range in Nanoose. During a slowdown at HYCO, he moved to Lockheed Petroleum where he stayed for almost 17 years, and then went on to Atlantis Submarines.

- **Tom Roberts** did quality control, purchasing and electrical work in HYCO's shop, then he switched to the field team, handling submersible maintenance, and eventually took on global piloting duties. He then moved to Atlantis Submarines and still does contract work.

- **James McFarlane Sr.** served in the Canadian Armed Forces for 18 years, reaching the rank of Lieutenant Commander. He retired in 1971 and became HYCO's vice-president of Engineering and Operations. He helped develop *Pisces IV* and *V* and coordinated getting *Pisces V* and its crew to the UK rescue site when *Pisces III* sank; that successful subsea rescue would become legendary and facilitated the acceptance of unmanned ROVS. James Sr. left HYCO in 1974 to found his own company International Submarine Engineering (ISE).

- **Fred Warwick** started as an electronics tech but quickly became a familiar *Pisces* pilot, training many who followed in his footsteps. He came and went with HYCO, working on various contracts.

- **Al Robinson** arrived at HYCO as a truck driver, a carpenter and an auto body worker. "I had grown up in a fairly normal world, but working for these guys was just mind-blowing. I'd never seen anything like this, but I was determined to be a part of it. I was hired as an assistant hydraulics technician, then they fired my boss and asked me to take over. I knew I needed more knowledge, so just persevered and that's how I learned."

- **George Hazelton** was an experienced hardhat diver who also ran the fibreglass shop at HYCO. Al Trice first met George when he was dive tender for Al Black at Burrard Drydock. "Blackie" taught them both hardhat diving and the use of explosives; in 1956, the trio worked salvaging logs for the Powell River Company.

- **Dennis Hurd** analyzed the technical feasibility for various HYCO builds and later took over running *Aquarius* and *Pisces* in the Gulf of Mexico, based at HYCO's office in Houston. He went on to found the successful Atlantis Submarines company.

HYCO also contracted with outside companies for specific needs such as batteries, penetrators, underwater communication systems, propulsion systems, cameras and other supplies. Les Ashdown states, "Because Don, Mack and Al formed HYCO here in BC, the expertise came here. Later, other companies spun off. But that was because all the expertise was here at the time. So if you wanted to work in the subsea industry, this is where you showed up."

Today, that offshoot legacy is a multi-million-dollar BC industry involving a hundred companies that can directly or indirectly trace their descent from HYCO, ISE or Can-Dive. While the commercial bandwidth for their products isn't wide in Canada, most companies regularly do business around the globe—especially with the US, China, Japan, Europe, Australia and New Zealand. Equally important, Canada is now fostering a third generation of engineers and technicians in the subsea workforce. That's quite a legacy!

Prior to building *Pisces 1*, Al Trice and Don Sorte had visited Chicago Bridge and Iron, the huge US company that had built the small submersible they had seen in California. The company wasn't interested in building any more submersibles, but Al and Don did meet John Horton, whose family had run CB&I for generations. Besides shipyards and other facilities, the company also had an underwater research lab, so it had some involvement in the subsea industry. A couple of years later, John came to visit HYCO, looking for investment opportunities. Al recalls picking him up at the airport and realizing John didn't have any socks on, which was unheard of back then. He'd just forgotten them, so the first thing to do was go and get him some socks.

At the time, Mack and Don were considering other options besides building submersibles. The tourist submarine business was one possibility, particularly operating in Hawaii. So HYCO took a second option to buy the *Auguste Piccard*, the sub that had successfully carried thousands of people to the bottom of Lake Lucerne during the Swiss National Exhibition in 1964. When the first buyer dropped his option to buy the sub, HYCO was suddenly next in line. "We'd already set up HYCO Maritime Industries as a joint effort with Horton and gone to Ottawa to apply for a government investment program with strict guidelines. But Horton decided he'd rather use *Piccard* in the oil business. And by then, we'd gotten really busy with *Pisces II* and *III*, so we dropped our option," Al recalls. Horton bought the *Piccard* in 1970, then later also bought the *Ben Franklin*. "We parted friends, and he spun off into Horton Maritime Industries" (see Chapter 11).

A new design for a new *Pisces*

After *Pisces 1* was launched, it quickly gained a reputation for the depth and breadth of work it could do. With each job, Al made notes of changes and refinements for future models. He detailed the practical side of a new design for these submersibles, and Boeing engineer Warren Joslyn drew up his ideas.

These included:

- bigger removable panels to facilitate getting at the electrical connectors
- a bigger sail (the piece that sticks up and forms a water-tight dam around the entrance hatch) that was removable for easy transport in an airplane
- a flat deck to facilitate walking around the sub's hatch
- a different skid (the part of the frame that supports the sub when on the bottom or on a ship's deck)
- a proper sonar

International Hydrodynamics had one significant ace in the hole as they began expanding their list of submersibles: they knew their vehicles intimately. Al Trice states proudly, "We not only built them, but we serviced them and operated them." That first-hand experience also translated into an awareness of construction problems to be solved. "*Pisces I* was a real maintenance headache," Al recalls. One particular problem was the steel penetrators (those pressure-proof holes that pass the electrical cables and piping into the hull), which rusted like crazy, necessitating gallons of WD-40. Covering the penetrators with a layer of stainless steel and then fibreglassing the hulls right up to the penetrators solved that problem. As a result, subsequent *Pisces* were much easier to maintain. A big plus of *Pisces I* and all future submersibles was that they could be moved easily by rail, road or air—a significant bonus as subsea work became global.

HYCO maintained a set of standard drawings for its *Pisces* submersibles. These drawings were for basic items such as the hull, tail sphere and trim spheres, the frame, skids and basic controls. The biggest difference in subsequent *Pisces* primarily related to their diving depth. *Pisces II* and *Pisces III* both had depth limits of 1,000 m (3,280 ft). *Pisces IV* was rated to 2,000 m (6,560 ft). Doubling depth capacity meant the hull's material, thickness and weight became significant factors. So the type of steel used varied, depending on depth requirements. Steel with the high-yield and high-impact strength sufficient to withstand pressure of 1,000 lbs per sq in (1,000 psi, equivalent to 70 kg per sq cm), which was required for a 2,000-m vehicle, was the grade that US Steel called HY 100 (other countries produced steel of equivalent strength

HYCO employee Tom Roberts explains the evolution of the *Pisces* class of submersibles: "*Pisces I* was the original prototype. If you look at several *Pisces*, you can see that the first one was quite different from the rest. From *Pisces II* onwards, the basic shape of the submersibles stayed the same." Photo by Gino Gemma, from personal collections of Al Trice, Terry Knight and Phil Nuytten

under different names). Valued for its strength-to-weight ratio, HY 100 became the outstanding steel of the US Navy's *Seawolf* class deep-diving submarines and was therefore considered classified by the US Department of Defense, a fact that created numerous problems for HYCO.

Tom Roberts adds, "Submersibles also change because from one year to the next, from one contract to the next, this whole industry evolves with newer, better, smaller parts. And the equipment on the subs—the sonars, lights, cameras, and so on—also improves. Every time HYCO built a submersible, we'd go with the latest thing, especially if it weighed less and used less power. We went from having to steal some company's directional gyro compass out of an airplane to using a Sperry marine compass that doesn't precess [drift] every time the sub makes a couple of turns. So the evolution of building subs just continues."

Pisces II and *III*—a new money pot!

In the late 1960s, Vickers Shipbuilding and Engineering in the UK decided to consider getting into the submersible business. The North Sea was just starting to show signs of oil; looking ahead they could see that supply boats and other equipment, including submersibles, would be needed. The Vickers shipyard at Barrow-in-Furness was operated by Len Redshaw. He realized that even though they were building nuclear submarines, it would take them forever to build one of these small submersibles. They had seen pictures, models, mockups and parts of submersibles, but no operational vehicles. Somebody at Vickers had seen a picture of *Pisces* in a champagne ad and gotten curious. Was it real? They contacted HYCO, looking to buy a small, operational, deep-diving submersible, and Redshaw sent two employees to investigate.

"We took them out to Deep Cove to see *Pisces I*," Al remembers. "They asked if it was a mockup. We asked who wanted to take the first dive! After that, they were indeed convinced ... and interested! Then we had to explain to them that *Pisces* was not for sale—but suddenly we realized that building submersibles could be our new money pot!"

Vickers invited HYCO to come visit their shipyard in Britain. "If we worked out a deal, they would pay our transport because we never had any money," Al explains. They put us up in a hotel and left us for a day to rest up. We'd never been to Europe before so didn't know anything about jet lag. We'd taken another Englishman, Al Hooper, with us. When we got to Len Redshaw's big office with its fancy woodwork, he looked at us, counted 'one, two, three' and said, 'Wait a minute, I've got to get two of my boys so we're even.'

"Soon we're sitting three guys on each side and Redshaw says, 'Ok, let's hear your spiel.' Mack and I had practised what we were going to say ahead of time. They listened while I talked. And Redshaw said, 'Oh, yes, yes, of course. Yeah. Yeah. Oh, you want us to fund you? Ok, that's no problem.' I can't remember how much money we asked for, but it was enough to fund *Pisces II* and *III*! After the deal was done, Len Redshaw told us, 'You should know that our board wanted to buy HYCO. But I put the kibosh on that, because if we bought you, you'd be Vickerized and we'd get nothing done.' Those were his exact words."

In 1968 HYCO began work on *Pisces II* for Vickers and *Pisces III* for themselves. Initially the plan for the new subs was that Vickers would fabricate two pressure spheres out of 1,000-psi steel, strong enough to allow them to go down to 2,000 m (6,560 ft). But when Vickers seemed to be taking forever to make the pressure spheres, Al decided that HYCO would build a 1,000-m (3,280 ft) vehicle out of a different steel instead. He talked to Victoria Machinery Depot, who thought they could compress inch-thick boiler plate steel using a powerful German steel press they'd gotten from the Department of War Reparations. "Initially, VMD was used to looser tolerances for lumber industry work and the initial six-segment hulls for HYCO were even out of round—no good for ocean work. But the company bit the bullet and learned how to make some for us that were plus or minus a few thousands of an inch."

When Al got back to Vickers with the new plan to build the 1,000-m vehicle locally, the UK company said that would be fine; they would take it. "So that's how *Pisces II* and *III* became 1,000-metre vehicles instead of the deeper-rated ones we'd originally planned." Al says proudly, "We built both those subs in eight months. We had 40 guys on the floor and our only management was a lady accountant and Bill Mackesy (who had worked with Al earlier in Hudson Bay) as company manager. *Pisces II* was delivered first and their crew trained in late 1968." Then HYCO finished off *Pisces III* and put it to work themselves. Years later, *Pisces III* also joined Vickers's fleet of submersibles.

Having been trained as a loftsman, Al Trice laid out and lofted the new submersible design full size. Then it was time to construct the framework for a plug, which is a careful mockup for the finished framework. Al searched fruitlessly for a fibreglass shop to do the job. Fortunately, he already knew quite a bit about fibreglass, so hired a couple of shipwright contacts to build the plug. Getting the spheres in the plug to be really round was the next problem, so he hired autobody workers. They came in at night, for extra pay, and did a perfect finish job.

The next step was making the mould on the plug, which involved coating the mould with a release compound so the resin wouldn't stick to it. Then several layers of fibreglass, each saturated with a resin, were added to a thickness of 3/16 in (5 mm). When this "skin," as it's called, is removed from the mould, it is strong, light, waterproof, and won't conduct electricity. HYCO used the same mould for every subsequent *Pisces* since the outer shape didn't change.

Al Robinson supplied the IDs that he could remember for this photo of HYCO's shop crew in 1968 (some of the spellings are uncertain). *Top:* Mack Thomson. *Upper row, left to right:* Al Robinson (part of the Switzerland crew); unknown person; Walter Fueg (the draftsman who went to Switzerland and later returned there); Mike Macdonald (in glasses); the next two men (striped shirt and sweater) were welders, possibly Stan and Gus; Bobbie Scribner from New Brunswick (v-neck); unknown person; unknown person (possibly a diver); Al Wheatley (wearing hat). *Lower row, left to right:* unknown person; Ron Alan; Rolf Glausser (another Swiss draftsman); Gino Gemma (purchasing agent, diver and NAUI instructor); Ken DeFoe/Dufault (white shirt); unnamed odd-job man (plaid shirt); Ray Cagna (machinist). Missing are Ken Waldie (also an in-house machinist) and George Hazelton (who worked with Ron Alan in the fibreglass shop).

Al Robinson remembered some other details. Ken was a French Canadian who hired Al as his assistant, but during a pressurized time at work Ken went to Reno to gamble; when he came home he was fired and Al took his place. The odd-job man also ran a small tug, and once when towing a *Pisces* he put it into a rock and it was laid up for a month. Al called Ray Cagna "a most incredible machinist." Ray joined HYCO a week after Al did and they worked together. Later, when Al went to ISE for four years, Ray worked there too. After Al came to Vancouver Island, Ray again worked with Al. Personal collection of Hugh Dasken, Al Trice, Al Robinson, Mike Macdonald and Phil Nuytten

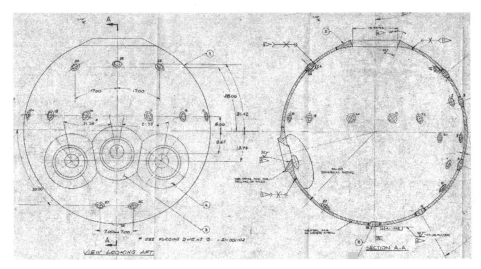

In an era long before computers, Boeing engineer Warren Joslyn made careful drawings for HYCO, beginning with *Pisces II*. This drawing is labelled "Main Hull (80") Weldment."
Personal archive of Deloye "Scratch" McDonald, with the aid of River Dolfi and Al Robinson

This learn-by-doing approach is all the more amazing considering that it happened in the era before computers. The stress analysis, all done by hand, was primarily Warren Joslyn's work. Having done the strain gauging on the B-52 wing when Boeing was building it, he was up to the pages and pages of calculations that it required.

That UK trip was the start of a fruitful ten-year relationship during which Vickers would purchase and/or lease six HYCO subs, as well as many spare parts. Vickers also spun off a subsidiary division called Vickers Oceanics as an operating company doing manned submersibles, with Peter Messervy as manager. The vehicles that resulted from the partnership between Vickers and HYCO, and the pilot training that Mike Macdonald and others provided, ushered in a new era and contributed to the establishment of a new wave of subsea operators in the North Sea.

A 1970 *Vancouver Sun* article quoted Don Sorte saying, "Our greatest single break has been getting connected with Vickers. It added credibility to our whole operation." Indeed, Vickers was a venerable shipbuilding company with a long heritage of building large naval warships and armaments. Now this huge UK company was very interested in tiny HYCO and would become HYCO's primary customer.

For years, Al Trice often worked side by side with Warren, just like an apprentice. "That's where I learned my engineering, and he was a hell of a good teacher." Sorting through a box of old binders, Al pulls out a thick one with pages listing the weight and balance computations for *Pisces*. "We had to collect all the info about every one of the items necessary for the sub—its weight, displacement, and where it would be placed on the sub. If the information was accurate, we could figure out crucial stuff like the vehicle's centre of buoyancy and its centre of gravity. Before computers, that computation would take days and days to compile. Now, with a computer, I can change any item on a weight and balance table and know the result in just minutes! But back then, you really had only one go at it because of all the time constraints and the cost."

HYCO Goes Public

In mid-1969, the original trio's partnership changed dramatically when HYCO became a public company. HYCO shares were listed on the Vancouver Stock Exchange, and the company's 1970 annual report included a note from Don Sorte saying, "As I am sure you are all aware, I was appointed President of International Hydrodynamics in May 1970." The same report detailed a subsidiary called HYCO International Engineering Inc., operating from San Pedro, CA. It was reportedly engaged in research and development of something called an "IFP Electrode." That project incurred a loss of over $71,000 and scuttled what would have been a modest net profit for the company's year-end report. There would be other projects, such as oil-eating chemicals and a coring device, which also didn't pan out. That HYCO subsidiary evidently continued to function in some capacity for several years with multiple vice-presidents but little to show for itself.

The move to a publicly owned company was controversial, particularly among HYCO's employees. Tom Roberts says, "I don't know whose idea it was to take the company public. I suspect it was Don Sorte's, maybe in hopes he could get more investors and have more money to develop things. Who knows where we'd have been today if HYCO had continued as it was."

Hoping to get involved with offshore oil and gas operations in the Gulf of Mexico, International Hydrodynamics also set up a subsidiary company in Houston called HYCO Subsea. Mike Macdonald recalls spending quite a bit of time there, sourcing jobs for Pisces and the Hudson Handler in the Gulf. "We did get some work, but it involved a lot of commuting."

SDL-1: Buying Local

In 1970 HYCO secured a contract with the Canadian Armed Forces to build a diver-lockout craft known as *SDL-1* (for *Submersible Diver Lock-Out 1*). Essentially the design was two pressure hulls connected by a short tunnel. The forward "command" sphere housed two pilots, with three divers in the aft sphere. At a maximum depth of 2,000 ft (610 m), the divers could leave and re-enter the aft sphere, return to the surface and connect with the ship's decompression chamber. Mike Macdonald recalls a pivotal meeting with Lieutenant Commander James McFarlane, who was tasked with procuring a suitable vessel. "He didn't know much about HYCO, so I took him on a success-ful dive in *Pisces*. He had been planning to buy a Perry submersible in Florida, but afterwards, he thought, 'Gee, Canadian guys can build one, too. To hell with the Americans; we'll get one from Vancouver.'"

As with many subsea projects, deadlines were tight for the *SDL-1*. Fortunately, HYCO learned that North American Rockwell Corp., one of the big aerospace companies, had built two hulls out of HY 100 steel for their *Beaver* submersible when they started getting into the subsea business. But the second submersible had never been built. So HYCO made their proposal based on using the spare hull and doing the design work themselves. The result was HYCO's first contract with the Canadian Navy. *SDL-1* was delivered in 1971.

Of course, building any government vessel was always more complicated. "There was a ton of paperwork," Tom Roberts recalls. "That's why I went back to the job of quality control technician for a while. *SDL-1* is also how International Hydrodynamics connected with Jim McFarlane Sr. When he resigned from the military, he joined HYCO and put his heart and soul into it."

One of the benefits of fabricating the *SDL-I* for the Canadian Armed Forces was that it connected HYCO with Lieutenant Commander James McFarlane Sr., who retired from the Canadian Navy and joined HYCO. Personal collection of Terry Knight

Nothing progressed without glitches, however. Also in 1971, Vickers Oceanics' new operations manager, Commander Peter Messervy, came to Canada for training in *Pisces III*. During a dive in Indian Arm, Cdr. Messervy and pilot Fred Warwick became stuck on the bottom 600 ft (180 m) down after someone omitted putting the ⅜-in (10-mm) vent plug in the tail sphere. Quite the initiation! Fortunately, the Canadian Navy's *SDL-I* was on the same barge undergoing sea trials. So with Mike Macdonald and Lieutenant Barry Ridgewell in the *SDL* and Al Trice on the surface, they extemporized some lifting lines to attach to *Pisces III*. Five and a half hours later it was finally recovered.

"When Messervy arrived back on deck, he was *not* happy," Al recalls. "I told him that we were just trying to make the training realistic!" Despite that incident, Vickers bought *Pisces III* for $490,000. This was the same Peter Messervy who later was instrumental in recovering *Pisces III* with pilot and co-pilot when they were trapped on the bottom of the North Sea for more than 80 hours. Al Trice was his right-hand man on that rescue. From their first encounter, the two shared a close friendship that lasted until Messervy's death in 2001.

Pisces IV, a contract that goes sideways, and *Pisces V*

HYCO heard that the Russians might be interested in a submersible. Al Trice vividly remembers: "They wanted to buy a little American submarine, but that was back when the Cold War was on, and Admiral [Hyman] Rickover said, 'No way. We're not selling any of our technology to the Russians.' So my Canadian Naval officer brother-in-law said, 'Well, why don't you write a letter to the Soviet Embassy in Ottawa and tell them you could build them a submarine?' So I did." In the meantime, HYCO went on to other things. But a year later, they got a reply. "I remember Don saying, 'Hey, we got a letter from the Russian Embassy.' And I said, 'Holy shit. They replied to my letter!'"

Al and Don Sorte went to Ottawa and secured an export permit for two Russian submersibles. "Ottawa wasn't all that concerned about security; they were just happy we would be creating work for Canadians." As negotiations got underway in 1971, Mack and Don then went to Moscow and soon figured out their hotel room was bugged. Meanwhile, Al went to Germany and talked to Rheinstahl, a big munitions maker in the Ruhr Valley, asking if they could forge spheres for two submersibles, with each hull consisting of two halves. They said that wouldn't be a problem, but they couldn't machine them. So Al arranged that the forged spheres would be sent to Switzerland for machining. Finally, plans were underway.

At the same time, Vickers finally came through with the 2,000-m rated sphere of high strength 1000-psi steel that they had promised HYCO long before. That deep-rated sphere could now be used for the Russians. HYCO dutifully got an export permit for it. Next Anatoly Sagalevich and Igor Mikhaltsev came from Moscow to see HYCO. Anatoly stayed as the on-site inspector and lived in North Vancouver; one of his sons was even born there. Al and Anatoly got to know each other well while working on the project and are friends to this day.

"I know Ottawa approved this deal, but has the Pentagon?"

This political cartoon ran in the *Vancouver Sun* on December 21, 1971, appropriately highlighting the question "Who's really in charge?" City of Vancouver Archives, Cartoon Scrapbook 1971–72 Temp 12-20-1971, cartoonist Roy Peterson

Near the end of the year it took to build a submarine for the Russians, somehow US Admiral Rickover got the power to cancel HYCO's Canadian export permit, which effectively negated the sale of *Pisces IV* to the Russians. At the time, Al was doing another survey up in Hudson Bay when Don got hold of him and said, "You'd better come back here." But he added that Fisheries and Oceans Canada (at that time part of the Department of the Environment) said they would buy *Pisces IV* instead.

Tom Roberts explains the problem that sank the Russian contract in 1972: "If you want to build a submersible that will go deeper, there's an automatic increase in the external pressure the hull will undergo, so you must upgrade to a stronger kind of steel like HY 100. But the US Navy was adamant that HY 100 steel was *their* military secret and there was no negotiating that point. That triggered the failure of the sale of *Pisces IV* to the Russians."

After the Russian fiasco, *Pisces IV* went to the Canadian Navy, then to Fisheries and Oceans, and finally to the Institute of Ocean Sciences, in Patricia Bay, near Victoria, where it did scientific research for several years (see Chapter 13). Meanwhile, HYCO completed *Pisces V* early in 1973, putting the sub to work burying underwater telecommunications cables off the coast of Nova Scotia. HYCO had secured part of a contract with the British Post Office to bury repeaters and splices of a new underwater telecommunications cable.

Terry Kerby, veteran submersible pilot and lead hand for the Hawaii Underwater Research Lab (HURL), states, "*Pisces IV* and *Pisces V* have been fine-tuned over decades to be the perfect science research submersibles. They made some major discoveries in terrain that would have been death to a tethered vehicle; hopefully those discoveries will have a positive impact on the management of these precious eco-systems. I love the *Pisces* submersibles. They are the perfect workhorses for these kinds of marine environments." Personal collection of Terry Kerby

The Canadians worked *Pisces V* on the Canadian side of the North Atlantic, and the Brits now had *Pisces III* on the European side.

That major contract was well underway when *Pisces III* sank off the coast of Ireland on August 29, 1973. That white-knuckle rescue focused worldwide attention on the vulnerability of manned submersibles as well as the emergence of remotely operated vehicles (ROVs) such as *CURV*. In the coming years, submersibles would be increasingly replaced by ROVs.

Pisces IV and *V* have had unexpected later careers. In 1985 Terry Kerby, an experienced submersible pilot with the Hawaii Underwater Research Lab (HURL), located *Pisces V* in a warehouse full of stockpiled submersibles in Scotland. The sub had worked in the North Sea for a while, then went to Intersub in France in 1979 and received a major reconfiguration that was not totally successful. *Pisces V* then moved to British Oceanics (successor to Vickers Oceanics), but that company soon laid up all their submersibles and relevant support equipment in a warehouse in Edinburgh. That's where Terry Kerby spotted PV in 1985. After an initial survey, *Pisces V* was purchased by HURL and shipped to Hawaii for a complete overhaul. Science dives commenced in 1987, and in 1999 *Pisces IV* also came to work for HURL. Two-sub diving operations started in 2000, all with a perfect safety record. Alas, in 2018 HURL lost its funding, their support ship was decommissioned, and the two hard-working research subs once again faced an unknown future.

Other HYCO problems

It's tempting to say that Don Sorte and Mack Thomson both left HYCO in 1972 "to pursue other interests," and let it go at that. But, as Al Trice notes, the truth was a bit more complex, if still lacking in details. "Mack had gotten himself into some serious legal difficulty selling stolen bonds to an RCMP undercover agent, so went to jail for 18 months. When I asked him about this, Mack wouldn't give any details. Don Sorte had grown increasingly dissatisfied with the way HYCO was going, resigning as president in 1972. He had his own hazy legal problems so he opted to take his sailboat and his common-law wife, Greta, and her two kids and slip away on an extended ocean voyage." That left Al Trice as the only one left of the original three founders.

Mack Thomson eventually got out of jail, but never rejoined HYCO. Instead he went to work for Terry Knight's electrical company. Terry says: "Later, Mack, Al Robinson, Neil and Flossie Wilson and a bunch of us started another little company called Canadian Underwater Vehicles. We even got a contract to build Russian submersibles, but we ran into big roadblocks and it went nowhere. Mack eventually moved to Guemes Island, WA, where he built a big shop and machined parts for the Atlantis tourist subs being built in Everett, WA. Around 2007 he moved from Guemes to a nursing home in nearby Anacortes, and some time later passed away.

Don Sorte's post-HYCO adventures were—and still are—the stuff of speculation. Terry Knight says, "Along with his other expensive toys, Don bought a yacht called *The Crusader*, but he was never an accomplished sailor. It was all he could do to get the boat anywhere close to a dock." Wisely, he hired crews that knew how to sail and navigate, though Greta also learned navigation. Al says, "A couple of years after taking off to sail around the world, Greta and her kids left the boat and went back to Australia. That pissed Don off."

Author Ken Dinsley, from Victoria, who had first interviewed Don, Mack and Al at HYCO, signed on as crew on *Crusader* sailing from Kenya to Singapore. Leaving Singapore, Don and other crew carried on, now heading back toward Sri Lanka. Al says, "Don used a ham radio and never bothered with the pesky business of a licence but used to talk regularly with a guy in Sri Lanka at a particular time. Who knows why he left for Sri Lanka when he did, with a typhoon brewing, but he did. He contacted the guy in Sri Lanka once more and said that they were in the midst of unbelievable seas, towing a big rope behind the sailboat, hoping it would serve as a sea anchor. But the next day at the regular time, he didn't call. And he never again communicated. So as time went on, various relatives of the crew started phoning me at HYCO. I didn't know anything but started realizing that all these people had disappeared." The boat and all aboard were reputed to be lost in a typhoon off India in 1977. Don was 47. Greta went to court to have him declared legally dead.

However, Les Ashdown tells an interesting variation on the tale that Don Sorte died at sea in the '70s. "Several years ago, Imogene, our secretary at Atlantis, ran into Don in Los Angeles; they even had lunch together! Then Don disappeared again. If he is still alive, he'd be in his late 80s or early 90s." While Terry Knight still believes that Don went down in that typhoon, he adds, "Knowing Don Sorte, nothing would surprise me."

Tracking down who took over the HYCO helm after Don Sorte proved a challenge. Fortunately, Doug Huntington knew that Richard "Dick" Oldaker might still be alive. He's now 93 and recuperating from a stroke but he spoke with his son Donald, who helped fill in some blanks. In late 1969, Oldaker took over managing the Vancouver-based Lockheed Offshore Petroleum Services, with his family joining him in 1970. *Pisces V* was still under construction when Al Trice and Don Sorte contacted him and convinced him to move over to HYCO in 1972, possibly earlier. Son Donald feels sure that his father put together the deal between P&O and HYCO. The motivation was twofold—to give P&O entry into the undersea services industry and to widen HYCO's customer base.

In 1973, the energy division of the P&O Steam Navigation Company bought into HYCO, forming a joint venture. Reputedly, they wanted a proper manager, "not one of these Don Sorte, Mack Thomson types," so Dick Oldaker became president. He had two MBAs and knew the oil industry—having worked for Standard of California and then Lockheed in Vancouver. Tom Rourke left P&O Subsea operations in the UK and came over to serve as HYCO's general manager. The new joint venture set up P&O Intersubs, a holding company for some of the *Pisces* submersibles, with offices in London and Halifax. The London-based company was owned 25 per cent by HYCO and 75 per cent by P&O; the reverse was true for the Halifax operation, with HYCO owning 75 per cent and P&O holding 25 per cent. The UK company also bought a 51 per cent controlling interest in HYCO itself. Doug Taylor notes, "HYCO had maybe a hundred employees, but now it had enough vice-presidents to rival Standard Oil." Indeed, a listing of officers showed that most executives carried *two* VP titles—one for HYCO and another for the mysterious California subsidiary HYCO International Engineering.

Imposing big company methods and office personnel on a small manufacturer like HYCO may have had some advantages, but it also stifled innovation and market savvy. Management also had difficulty adapting to new technology and to ideas such as remotely operated vehicles. They were determined to stay with what had worked in the past—manned submersibles.

Delving further into the company's economic difficulties, Dennis Hurd notes, "The problem was that the new management of HYCO really didn't understand that you couldn't make a company out of just building submersibles. You need a lot of people to do that. And one year you might build two of them, but then not build anything the next year. We could probably have made HYCO work if we'd kept focused on operations."

Slowly, things began to unravel in the company, though in many respects, it was still business as usual. A number of HYCO's frustrated employees jumped ship, moving out to form their own companies or work elsewhere. "Generally, I don't think people realized that HYCO was sliding into bankruptcy," Dennis Hurd says. "Or maybe they just didn't want to believe it. When P&O got involved, there was some idea that they might save the company. But that didn't happen either."

Aquarius, the shallower, cheaper alternative, 1973

Al Trice remembers *Aquarius* as Mack's dream: "It didn't work worth a shit to start with. Mack was an artist and an inventor with no sense of weights and balances, so we had to add a lot of foam. I didn't call it a submersible; I called it a happening." Personal collection of Mike Macdonald

HYCO's next submersible build was the smaller *Aquarius*, complete with a 40-in (101-cm) acrylic dome and less depth rating as a result of the big window. "*Aquarius* just didn't have the payload capacity to be commercially viable," Tom Roberts recalls. However, once the kinks were worked out, HYCO Subsea did find work for *Aquarius*, initially as an observation submersible doing platform inspections and pipeline surveys in the Gulf of Mexico. Tom notes, "We mounted a sensor on the sub so we could run along each pipeline and see if it was rusting away or not."

The remarkable *Pisces VI*

The question still remained, what to do for the Russians? Al Trice explains: "Back then we used to copy all our blueprints and drawings onto microfilm because there were no computers yet. Originally, the plan was that the hulls for what would become HYCO's *Pisces VI* were to be for the Russians. As a workaround to building in Canada, our new president Dick Oldaker hustled off to Switzerland and deposited that microfilm of our drawings in a Swiss bank vault. Then he said to Sulzer, the big Swiss machine works based in Winterthur, 'Do you mind if we build a submarine in your country?' He assured them we'd hire local workers, so they agreed. That's how the Swiss built what we thought would be a hull for the Russians from the Rheinstahl forgings. But the Swiss took forever. I guess they thought they were building a watch. However, when the job was finally done, it was perfect. The spheres from Switzerland were machined to three thousands of an inch! So, instead of using them for the Russians, we decided to keep those hulls for HYCO. In 1976, that sub became *Pisces VI* and set a number of records for deep-diving work."

Tom Roberts agrees: "*Pisces VI* was an amazing piece of equipment. The hulls for almost all HYCO-made vehicle were sent down to the Southwest Research Institute labs in San Antonio, Texas, for pressure testing. Using a very large pressure chamber, they can squeeze hulls, simulating great depths. *Pisces VI* was built for a 2,000-metre depth capability, which is a mile and a quarter straight down. The lab took them to the full pressure rating of their test chamber and there was still no sign of yield whatsoever. In fact, the American Bureau of Shipping (ABS) gave the sub a deeper depth rating of 2,500 metres because of the excellent pressure test results. Everyone who ever dived in *Pisces VI* felt secure" (see more in Chapter 14).

Finally, submersibles for Russia:
Pisces VII and *XI* (1975 and 1976)

After the earlier forced cancellation of the sale of *Pisces IV* to the Russians, a new era of détente between Russia and the West in the early 1970s helped renew and facilitate contact with the Soviets. They still wanted a deep-diving research submersible rated to 2,000 m (6,560 ft). In 1974 Doug Taylor joined HYCO to manage the project. Initially, the new contract, dated September 23, 1973, was for $3.85 million. It specified a *Pisces* and a diver-lockout submersible, spares, training and sophisticated high-tech instrumentation. That contract was quickly amended to be two *Pisces*, spares, training and the instrumentation, as well as severe late-delivery penalties for failing to meet the May 1, 1975, and May 1, 1976, delivery dates.

Vickers Shipbuilding in the UK had finally produced the hull spheres that they had promised years ago—now slated for *Pisces VII*. They were fabricated from high-strength 1,000-psi steel and had cleared static pressure tests to 2,000 m. HYCO was concerned about bringing what were essentially HY 100 hulls back to Canada on the chance that the US Navy might again pressure for another contract cancellation. So the Vickers spheres from England were tested in Marseille, the submersible was assembled in Switzerland, then went for sea trials to Italy, where the Russian crew was trained as well—*Pisces VII* was truly international. Building the initial Russian sub in Switzerland meant HYCO didn't have to worry about an export permit from the Canadian government—or American interference.

A small, select HYCO crew became international as well, going over to Winterthur, Switzerland, for many months of specialized *Pisces* assembly. Walter Fueg, himself originally from Switzerland, was part of that team. He spoke German, arranged for apartments, sometimes took the crew out to restaurants to break up the boredom of their own cooking, and did a lot of the gofer work that made the job comfortable.

Doug Taylor calculates that fabricating a *Pisces* submersible generally entailed about 15,200 man-hours of manufacturing labour and about 5,000 hours of engineering labour. Based on 2,000 regular man-hours per year, an eight-man team could complete a *Pisces* in about a year. However, the international crew were experts in *Pisces* assembly, loved their work, and actually got the assembly done considerably faster than in Vancouver. In 1975, the first Russian submersible, *Pisces VII*, was loaded aboard a Russian freighter in Savona, Italy, and headed to its new home in the Soviet Union. And the critical final payment for it was received.

But the Russians had contracted for two subs. Doug Taylor was well aware that costs would increase significantly if a second sub was built in Switzerland. As well, given previous delays, he had concerns about delivery delay penalties. Al recalls, "For the second sub, we went to the Canadian government and said, 'The Russians want another one. Do you want us to build that one in Switzerland, too?' 'Oh, no, no. Build it here in Canada.' That solved that problem."

However, the contract for the second sub also stipulated delivery of many advanced instrumentation systems. None of that instrumentation was for *Pisces XI*; in fact, many of the instrumentation systems were highly sensitive and not allowed under Canada's export control legislation. The Russians had hoped to sneak the instrumentation in as part of the submersible contract, a ruse that created many headaches for HYCO. To try to rectify this problem, Doug Taylor and HYCO president Dick Oldaker travelled to Moscow in late October, with a first stop in Ottawa for Doug to meet with Canada's Export Development Corporation (EDC). Nothing about that trip to Moscow was straightforward or helpful, Doug says. "There were a number of meetings but all were totally unproductive. The instrumentation problem wasn't resolved until 1975, when the Russians finally agreed to buy the specialized equipment separately from Norway." The contract price was significantly reduced to $2.65 million.

Yet another problem loomed: where to get the second set of hulls with a 2,000-m rating. At the time, HYCO had contracted with Victoria Machinery Depot (VMD) to build the hulls for *Pisces VIII* and *X*, each with a 1,000-m rating. But that depth was only half of what the Russians wanted. Jim Ferguson was project manager on one of the Vickers contracts and had a background in shipbuilding. He heard that VMD said they could build 2,000-m rated hulls at their shop using Welten 80C steel if they made the hulls out of twelve smaller segments instead of the six larger ones normally used. Furthermore, the cost savings to the contract would be more than $200,000. Additionally, they could be completed faster, thus avoiding the significant delivery penalties that had Russian bureaucrats eagerly wringing their hands in anticipation.

Al says, "It was a hell of a gamble, but what choice did we have? The *Pisces XI* fabrication was approved, and the hulls were completed at VMD. The final hurdle was pressure-testing that 12-segment hull, which we did again at the Southwest Research Institute. By god, it pressure-tested just fine, and we met the May 1, 1976, deadline. Needless to say, the Russian bureaucrats were totally pissed off at missing out on the penalty money."

But there was still one more challenge, and that was telling the Russians that HYCO was substituting the *Pisces XI* hulls for the *Pisces VI* ones. They were definitely *not* happy. In fact, the swap almost blew up the contract and there was huge push-back for months. Even getting the Russians to attend sea trials for *Pisces XI* was an ordeal. Doug Taylor vividly recalls the night-before-sea trials session with the Russians and the challenging diplomacy that involved gifts, much vodka and pickled herring. It worked, and sea trials were completed on time. The second submersible was shipped out of Vancouver on a Russian freighter and payment was made promptly. That's how Russia was finally able to secure two deep-diving *Pisces*-class submersibles, *VII* and *XI*, for use by the Shirshov Institute of Oceanology of the Russian Academy of Sciences.

Once delivered to the USSR, Anatoly Sagalevich was in charge of operating the *Pisces* submersibles. Al says, "The Russians had a big research ship,

This 1990 Russian stamp commemorated the work of the *Pisces* submersibles in the Soviet Union. Personal collection of Al Robinson

the *Akademik Keldysh*, and, boy, for the next ten years Anatoly ran those subs right to the limit—2,000 metres. Eventually, he came up with the idea of designing new submersibles that could go clear down to the *Titanic*. Somehow the Russian government came up with enough money to build two *Mir* submersibles in Finland. When those were operative, one *Pisces* went to a museum in Kaliningrad and parts from the other were used as spares for the *Mirs*."

With the Russian contracts successfully completed, Doug Taylor transferred to HYCO Subsea operations. He left the company two years later after negotiating a significant contract with Getty Oil consortium for *Pisces VI* to operate from the big drillship *Discoverer Seven Seas*, off Pointe Noire of the French Congo. When launched in 1975, the *Discoverer Seven Seas* was the most sophisticated deepwater drillship of its time. Keeping a ship exactly in position is crucial for a successful drilling operation. To do that, the *Discoverer Seven Seas* was dynamically positioned so instead of relying on massive anchors to try to maintain position, it utilized a satellite system for surface positioning, an acoustic beacon system deployed on the seafloor and a set of very large omni-directional thrusters to maintain a fixed position relative to the sea floor.

Getty Oil had acquired exploration rights off the coast of the Congo. The water depth of the planned site was 4,752 ft (1,448 m), the deepest that the *Seven Seas* had ever operated. Al Trice, Teddy Agon and Tom Roberts followed *Pisces VI* to Africa. Al recalls, "The Brazzaville Congo, the old French Congo, was a total Communist country and very dicey. So Getty asked us, 'What's your submarine worth?' We told him a million dollars. 'OK,' he said, 'we'll write it into your contract so if they steal your sub, we'll give you a million dollars.'" *Pisces VI* pilots had to guide the very long drill stem hanging from the ship into the conical section above the two-storey-high stack of blow-out preventer valves (the BOP stack). This occurred during re-entry operations whenever sampling or drill bit changes were needed. On site, Al and Tom made the first dive—at 4,400 ft (1,340 m), a record depth in oil and gas work. At the time, these were also the deepest commercial dives made by any *Pisces.*

Years later, International Underwater Contractors, an American diving company, purchased *Pisces VI* for rig support in the Hibernia oilfields off Newfoundland. Despite continuing to establish new diving records, it was eventually retired and put in storage when ROVs became more viable for the oil and gas industry. Then, in 2015, *Pisces VI* got a reprieve when 30-year-old Scott Waters, from Salina, Kansas, bought the submersible. The owners accepted his low bid because he presented a serious plan for refitting the sub and putting it back to work. Once that refit is complete, *Pisces VI* is slated for underwater scientific research (see more in Chapter 14).

Doug Taylor negotiated HYCO's first significant contract with the Getty Oil consortium for *Pisces VI* to operate from the big drillship *Discoverer Seven Seas* in the Congo. Dennis Hurd then negotiated a second contract with them for HYCO; the drillship, along with *Pisces VI*, came to the Mediterranean. He's shown here with *Pisces VI* on the deck of the *Discoverer Seven Seas*. Personal collection of Dennis Hurd

Pisces VIII and *X, Leo* and *Taurus*

In 1976 HYCO had more than 165 people on the payroll. But a surplus of submersibles was slowly building up worldwide as ROVs became more popular. And no one could foresee the drastic cutback of Britain's North Sea operations in the late 1970s. HYCO had completed *Pisces VIII* and *X* for Vickers Oceanics in 1975. They wanted a big window for better visibility, but that meant both subs could only be rated to 1,000 m (3,280 ft). That depth was sufficient for the North Sea (which is nowhere deeper than 700 m / 2,300 ft) but was prohibitive for other, deeper-diving jobs. P&O Subsea put in the company's next order, for *Leo*, another submersible with a large window and hence a limited depth rating. It was built in 1976 and also intended for North Sea work. Tom Roberts does not mince words: "*Leo* was the stupidest sub I ever saw. Keeping that thing running was a pain."

Pisces IX was slated to be built and HYCO even had the forgings for it in Germany, but the market for submersibles dried up, so it was never fabricated. That left *Pisces VI* as probably the last HYCO vessel that was truly commercially viable.

As technology evolved in the oil and gas industry, the trick for HYCO, or any subsea company, was trying to stay ahead of changing trends. In the mid-1970s, many companies were using saturation divers for wellhead work. Lockheed Petroleum already had a device based on the McCann Rescue Chamber, a kind of diving bell, but it was tied to the surface. Al Trice thought about this and figured there might be a market for a tough submersible that could carry heavy gear, do a dry transfer of personnel for maintenance at one atmosphere pressure on a Lockheed wellhead at 2,000 ft (610 m). It could also do diver lockout at 1,000-ft depth (305 m). His idea became *Taurus*, and Warren Joslyn did the design. Plans changed in 1976 because of a three-year wait for HY 100 steel. Instead, Vidor Engineering, in Newcastle, Australia, forged the hulls of boiler plate steel with a limited 1,100 ft (335 m) rating. They were shipped back for pressure testing and HYCO completed the new vehicle.

The fact that HYCO president Oldaker had formerly worked for Lockheed influenced the development of *Taurus*. Tom Roberts adds, "We were in bed with P&O and they liked the idea, so financed it. The Canadian government put in some money as well. So eventually, in 1977, we got the new vessel built, tested and it worked just fine." Things were looking good for *Taurus*, particularly when Vickers chartered it to use in the North Sea for diver lockout. But then two drastic things happened in the late '70s. First, the British government nationalized all shipbuilding in 1977. Suddenly, Vickers Oceanics' submersible operations no longer existed. Second, P&O had a change of management. At the time, they had the largest supply boat fleet for servicing offshore rigs, but the new boss decided to get out of that business. So P&O got rid of their supply boat fleet *and* closed its HYCO connection. *Taurus* was left behind in Scotland, the last underwater vehicle that HYCO would ever build.

Alan Whitfield, former senior pilot at Vickers Oceanics and the Norwegian J/V Fred Olsen Oceanics, has provided information on the sub's later career. He is the owner of Silvercrest Submarines, a UK company that sells or rents submarines and submersibles. In 1992 the company purchased and refit *Taurus*, then completed more than a thousand dives to 800 ft (245 m) in Loch Ness (monster hunting) and in the South Atlantic off Cape Town. In June 2020 *Taurus* was shipped from South Africa to the US for a Department of Defense project.

In 1978, P&O announced they were closing HYCO. An article in *The Province* of September 29 noted that trading in shares of International Hydrodynamics Ltd. on the Vancouver Stock Exchange had been halted at the request of new company president Cliff Sawyer. It also mentioned that a general meeting had failed to resolve a conflict between P&O management and dissident shareholders who owned more than 30 per cent of HYCO's stock. Further along, the article noted that the dissident shareholders were led by former HYCO president Richard Oldaker. Ultimately, P&O and the holdout shareholders failed to agree on a rescue plan, effectively sinking HYCO.

With all hope of rescue gone, the receivers came in to HYCO with the purpose of winding down the business. Former HYCO engineer John Witney recalls: "They weren't even trying to keep it going and let everybody know that." He adds that later they stopped by again and were obviously puzzled. "Why is everybody still here? Why aren't you out looking for other jobs?" The reply was unexpected: "We don't want another job. We've worked so hard here for so many years that we're just going to take a couple of months off." John adds, "Well, that mindset was rather new to the receivers, but it tells you that HYCO was a great place to work. Absolutely."

After laying off most of the staff, the receivers tasked Doug Huntington with sales in hopes of keeping the company running until a buyer could be found. They also said that they'd continue to operate anything still making money. HYCO had a few active contracts, so those continued until completion. One contract was with NASA. It involved recovering a space capsule's large booster rockets after they were jettisoned at altitude and parachuted down into the ocean, where they flooded and settled vertically. NASA already had an ROV capable of locating the rocket engines, so Al Trice designed a way to attach an air hose to the ROV. It then blew air into each booster fairing, forcing the water out so that it floated horizontally and could easily be towed to shore for recovery. Al says, "It was an interesting, successful project where space and underwater technology worked hand-in-hand."

By 1979, all contracts were completed, leaving Dennis Hurd and Al Trice as the last remaining HYCO staff. "I was responsible for the drillship business, and Al was the main technical guy. Because the bank wanted to get some money out of the company, they proposed that we put in a bid to take over the submersible and drillship business. Al and I did that and won the bid. But being the little devils that they were, the receivers thought they could do better than that so announced they were going to have a second round of bidding," Dennis explains. "This time, Comex got involved and beat us out of it. That was the end of the HYCO days for all of us."

Over a fifteen-year period, HYCO built fourteen submersibles, the deepest-diving being *Pisces VI* with a depth rating of 2,500 m (8,200 ft) and the largest being *Taurus*, weighing in at 54,000 lbs (24.5 tonnes). "Not only did we design, build and sell submersibles, we operated them," Al Trice says proudly. "This operational experience fed back into the design/build loop and that's what kept us in the manned submersible lead."

Al's commitment to HYCO began in 1964 and ended in 1979. It spanned the heyday of deep-diving manned submersibles until their replacement by unmanned robots. By this time, Vancouver was already on the map as a world leader in the subsea arena. Diver, inventor and subsea businessman Phil Nuytten states, "It's a little-known fact that much of what is now routine deep-sea diving stuff worldwide was actually developed right here in BC."

Indeed, it's difficult to quantify that impact, since specific statistics on the subsea industry, nationally or provincially, are hard to come by. In 1978 a federally sponsored sector task force identified 180 companies in Canada's ocean industry, with more than 50 per cent of their sales and subsea services exported. By 1990, a report by the SPARK OCEANS committee of the Science Council of British Columbia noted 62 companies in BC's marine robotics and subsea vehicle industry alone. With annual sales of $62 million, "this industry plays a significant role in moving Canada from a resource-based to a knowledge-based economy." The marine electronics sector also employed more than 500 people in the province and added a further $40 million to that figure. Perhaps the most telling comment on this little-known industry came in a 2007 report put out by the Canada/British Columbia Oceans Coordinating Committee. It admitted quite honestly that the general ocean-based economy of BC is "much larger than previously estimated, more diversified than previously thought, and not well understood."

In his May 2000 "On the Waterfront" column in *Harbour & Shipping* magazine, James Delgado summarized: "British Columbia is an international centre for deep ocean research, with a number of multi-million-dollar businesses actively engaged in the exploration of earth's final frontier." He notes that *Pisces I* was the "genesis of a new, made-in-BC competitor in the global submersible and deep ocean market." He then quotes author Tom Henry: "Although HYCO folded as a business in 1979, its subs, in the hands of other firms, continue to work to this day, as do its former employees, many of whom now manage Vancouver's elite underwater research firms. Among this crowd, HYCO is referred to as the 'kindergarten'—the place where it all began."

CHAPTER 5

Phil Nuytten: Can-Dive, Oceaneering and Nuytco

1966–present day

———

S tarting in the 1960s, Phil Nuytten built a reputation that centred on a "can do anything underwater" work ethic. He helped pioneer deep diving and the development of mixed-gas decompression tables and then diving in icy polar waters. Later, he would become known globally for his atmospheric dive suits and small scientific submersibles.

With a larger-than-life personality, Phil also attracts a variety of opinions, ranging from "pirate" and "bullshitter" to "true genius." International Submarine Engineering founder James McFarlane Sr. calls Phil "a national treasure, a pioneer of substance." James A.R. McFarlane Jr. adds his perspective: "Phil and my father have been great friends, although they were fundamentally going at the underwater world from different directions. My dad was military and educated in engineering, whereas Phil is a very talented artisan. The difference is that my dad is an engineer's engineer—he's totally binary. Phil is an artist's engineer and builds beautiful vehicles." Certainly, many in the subsea industry would echo Jim English, a former employee of Phil's, who states, "Phil opened most of the doors, windows, hatches and ice holes that became my career."

———

By 1966, at the age of 25, Phil had saved enough money from non-stop commercial diving to start his own underwater diving construction firm. Originally, he called it Industrial Marine Divers, later changing the name to Canadian

Divers. Soon that shortened to Can-Dive, and Phil set about becoming one of Canada's largest underwater contractors. Initially, he recruited fellow union members on a per-job basis. It took a while before he could afford to put anyone on the payroll.

The challenge of underwater construction work

Underwater construction contracts typically range from inspections and repair to installations or salvage. To bid and complete any of these jobs, a company has to own, rent or borrow

From the beginning, Phil Nuytten knew that underwater construction would be Can-Dive's bread and butter. Job conditions are challenging, often cold, dark, and in water that is deep, has current and tides to deal with, or has zero "viz" (visibility). Often one job has them all. Personal collection of Phil Nuytten

an arsenal of "underwater intervention methods"— that's contractor-talk for the specialized air and gear any marine job might require. Furthermore, that equipment has to get to and from the site as needed. That means barges, tugs, support ships, and workboats, or sometimes trucks and planes.

The commercial divers who execute this work and run the equipment are generally skilled in several methods of diving. If relying on air supplied from the surface, they use either traditional heavy copper/brass helmets or the lighter-weight fibreglass Ratcliffe helmets, known as "Rat Hats." Deeper jobs require divers to use complex and expensive diving options like helium/oxygen diving, bell/bounce diving or saturation diving. If the job is not suited for a regular diving intervention, then the company must supply a remotely operated vehicle (ROV) with manipulator arms or an atmospheric diving system (ADS), or even a submersible, all of which have high costs.

As Can-Dive acquired steady work, the company developed a core team. Norm MacDonald was Can-Dive's lead diver and supervisor, spending more time working underwater than most people did in their day jobs. Overall management was handled by Dave Porter and Don Romaniuk. Phil was involved in almost every job. Don Leo Jonathan, the famed heavyweight wrestler Phil had met back in his early dive shop days, became a partner in Can-Dive. "Don Leo might seem like a strange choice for a diving business partner," Phil admits, "but he was a class act, and his word was his bond."

Phil remembers Don Leo saying, "What could be less important than last Monday's televised wrestling match? I want to do something with my hands that will last." So Phil invited Don Leo to join his crew on a job in Hawaii, redoing bridge foundations. "He loved that and became a life-long friend and my eventual business partner. Whenever Don Leo wasn't wrestling somewhere else in the world, he'd join in on projects. Phil and Mary became close friends with Don Leo and his wife, Rose. "Don Leo was away wrestling in Winnipeg when his youngest son Jeff was born, so I used to kid him that my wife and I met his son even before he did." Today, Jeff is all grown up and is Phil's second-in-command and lead sub-pilot at Nuytco Research.

From its early years, Can-Dive followed Phil's simple work ethic:

- Get the job done safely at all costs—never compromise safety.
- Keep your word to the customer and your mates. Do whatever you said you were going to do.
- Meet the schedule.

Missing from that list is "do it on budget." Of course, like any company, Can-Dive always has to deal with a contractual budget, so the client is consulted if unexpected costs arise.

Aside from equipment, diving and underwater ops are all about people and personality management. Keeping the troops happy while maintaining a professional attitude and discipline can be a challenge given the very strong, independent personalities that divers are known for. In the early years, a lot of this personal interaction took place at local pubs over very long lunches.

Project management primarily involved just applying common sense and then staying awake nights to worry about details and not missing anything. "Phil was the dreamer and creative leader, with great ideas and concepts for inventions and ingenious ways of doing a particular job," Doug Elsey and Jim English agree. "Then it was up to us, and folks like Norm MacDonald and Dave Porter, to actually turn the idea into something that fit with the hardware, people and job completion objectives."

Hooking up with Lad Handelman and Cal Dive

In the mid-1960s, Shell Oil Canada was set up to begin deepwater drilling in Western Canada. The new oil company dangled a tempting five-year contract for any service company with deep-diving/bell-diving capability. Major companies like Ocean Systems, Comex and others were all salivating over such a long-term strategically located contract. What chance did an upstart company like Can-Dive have, particularly with no diving bell and no helium capability? Likely none.

Phil Nuytten was well aware that he had neither the equipment nor the experience to qualify for the contract. Nonetheless, he went to Calgary to talk with Shell Canada. "Can you work as deep as 600 feet?" they asked. Phil had to admit he couldn't but was willing to learn. Shell gave him a firm "No."

Undaunted, Phil headed to California to see if he could cut a deal with a large diving company like Ocean Systems. They, too, told him to forget about it. In fact, they were intending to bid for the job themselves. Phil recalls: "I headed out, thinking it was hopeless. But one old-timer followed me out into the parking lot and slipped me a paper with a phone number, adding, 'That's my nephew. He's got oilfield experience.'" The nephew turned out to be Lad Handelman; not only was he a helium diver but he also had a track record working for the oil industry—exactly what Phil needed.

In fact, Phil and Lad were remarkably alike. Both were risk-takers and hard workers. Both had become commercial divers when young, pushing themselves to greater and greater depths. Both had formed small commercial dive companies with similar names—Canadian Divers (Can-Dive) and California Divers (Cal Dive). And both young companies were gaining traction, though in different areas. What Lad lacked in construction know-how, Phil had plenty of. In return, Phil hoped to learn from Lad's helium diving experience in the oilfields.

Nobody expected Phil Nuytten to join Lad Handelman to form a 50/50 working partnership. After a memorable presentation to Shell, and against all odds, Can-Dive won the prize—a five-year contract with Shell Canada.

As a result, Phil began to amass oilfield experience, and Can-Dive gained extensive exposure to the oil industry. Beginning in 1966, Can-Dive successfully provided diving services for Shell's first offshore oil well in Canada, drilling in 600 ft of water (180 m) in various locations off Vancouver Island. That work took place on the SEDCO 135-F (the same rig on which Mike Macdonald and *Pisces* would work in 1969—see Chapter 4). This was the first drilling platform built in British Columbia, and in 1966 it was chartered by Shell Oil exploration. The SEDCO was hard to miss, floating 50 m (165 ft) high on the water. Over a period of two years, fourteen test wells were drilled off the BC coastline. Shell would eventually abandon its offshore drilling program in BC after spending $33 million and finding no significant oil or gas deposits.

For Can-Dive, the SEDCO work typically involved 8-hour dive shifts, although divers were essentially on call. Three divers would squeeze into the bell chamber that was lowered from SEDCO's spider deck (lowest deck) to the drill site on the seabed. Two were dressed divers, with the working diver outside the bell on the well site and a fully dressed standby diver inside the bell, ready to render assistance instantly if needed. The third man in the bell was the tender, who was in constant communication with the surface operator on the spider deck. He also continually monitored and regulated the supply of oxy/helium or air that the working diver was using.

The Shell Canada contract essentially made Can-Dive Services Ltd., as it was known back then. But work on the SEDCO 135-F was often in horrendous weather conditions. If it was too dangerous, crews might be transferred to the SEDNETH-701, a drill rig that was designed as a SEDCO 700 series and built in 1973 by Hawker Siddeley Canada Ltd. of Halifax. At the time, this alternative drill rig was working in West Coast waters, but later on returned to Atlantic Canada. As a result, Phil opened up offices in Halifax and St. John's, conducting other diving work on Canada's East Coast when not doing oil work.

In the coming years, Can-Dive would do similar work for Amoco off the coast of Newfoundland, for Pemex in Mexico, Petrobras in Brazil, and Shell, Occidental and BP in the North Sea. From 1975 to 1987 Dome Petroleum was drilling in the Beaufort Sea, with Can-Dive providing drill support services in the High Arctic. That work encouraged Phil to come up with a variety of sub-zero diving innovations such as gas-heating equipment and heated suits.

Although Lad Handelman's Cal Dive had many successes of its own, notably in the lucrative Southeast Asia market, his company still had no presence in the Gulf of Mexico. But Lad was well aware that big opportunities were about to open up in the Gulf, with the oil and gas industry increasing its spending there from $3 billion to $25 billion. In order to keep pace with major competitors in the Gulf, Lad's company needed capital to build diving bells and other expensive deep-diving systems. But cash was in short supply. Major industrial corporations were busily buying up small diving companies, a temptation that became very real for Cal Dive when they received a similar buyout offer. They were given 45 days to decide whether or not to accept a deal that would provide cash and security.

Phil Nuytten cornered Lad and reminded him why they had

Phil Nuytten on the spider (lowest) deck of the *SEDCO 135-F* drill rig. He's conducting minute pre-dive checks on the manifold that controls the diver's oxy/helium or air mixture. The surface operator also monitors the umbilical that includes delivery of air or mixed gas to the working diver. The decompression chamber behind Phil shows the opening where the bell locks onto the decompression chamber, allowing the crew to transfer from the bell and begin decompression. Personal collection of Phil Nuytten

joined together in the first place: "Not to have idiot bosses to report to and to be able to do our own thing the way we want." Lad agreed, then signed up for a seminar on mergers and acquisitions. By chance, that's where he met Matthew

Simmons, a young post-grad researcher at Harvard. Hearing of Cal Dive's buy-out offer, Simmons countered with one of his own. Within the remaining 30 days, he would bring a firm offer of funding for which Cal Dive would only have to give him a minority position, so they could continue to be their own boss. And that's just what Simmons did—he traded $350,000 in financial backing for a 30 per cent stake in the new company.

Oceaneering emerges on scene

In the summer of 1969, Lad Handelman and Phil Nuytten co-founded Oceaneering International, registering the company in tax-favourable Delaware. Months later, in December, they bought out World Wide Divers of Texas in return for Oceaneering stock. Major stockholders in the new company now included Lad Handelman, Phil Nuytten, Mike Hughes, Johnny Johnson, Bob Ratcliffe (inventor of the "Rat Hat" diving helmet), Frank Stolz and Kevin Lengyel. When Peter Barbara, a young insurance broker, stepped in, offering to provide necessary insurance coverage, the contracts for the fledgling company began to roll in. Using a "stock for pay" strategy, Oceaneering attracted new expertise, particularly in everyday financial control and executive leadership.

Can-Dive kept its own name but functioned as a subsidiary, calling itself Can-Dive Oceaneering. Phil was a board member and senior VP of Oceaneering; he also continued as president of Can-Dive, with his company responsible for all marine operations in Canada and the North. Can-Dive was the largest diving company in Canada, with offices, at various times, in North Vancouver, Tuktoyaktuk, Mississauga, Montreal, Halifax and St. John's. A significant chunk of emerging work was support services for the oil and gas companies in the Arctic and off the East Coast. Can-Dive operated several saturation dive systems in those areas, using manned submersibles and ROVs. "Those years were really tough on family," Phil recalls. "My daughter Virginia was very young, but my job with Oceaneering demanded travel. I was developing new diving systems but also opening new territory like Brazil, so I wrote her letters from all over the world."

Phil laughs, recalling an early board meeting after Oceaneering was formed: "Each of the directors got up to say, 'OK, this is what I want to do with this company, this is where I think we should go.' When it was my turn, I said, 'Atmospheric diving suits.' This was way before the JIM suit or anything else. But I had read about the Italian Galeazzi diving suit and was fascinated by the idea. Everybody on the board sort of patted me on the head and said, 'Phil, for Christ sakes, we're in the business of doing deep helium diving and saturation diving. Why would we build an atmospheric diving system that would compete with what we do?'"

Board members added that no one had been able to successfully make such a suit anyway. Phil replied that he thought he could. To which they said, "Well, if anyone could do it, you probably can, but why the hell would we let you do something that would put us out of business?" Phil responded, "If we don't do it and someone else does, then *they* will put us out of business." But nobody bought that argument.

About a year later, the magazine *Offshore*, produced by the subsea services firm American Oilfield Divers, published an article about the JIM suit that had just been developed in the UK. Suddenly Phil's phone rang off the desk with calls from fellow board members. They convened an emergency meeting to figure out what to do. Phil's advice was straightforward: "You should authorize me to go to the UK and test the suit. If it works, then we should cut a deal with the owners or try to buy them out."

Phil and one other board member flew to England to meet Mike Humphrey and Mike Borrow, partners in Underwater Marine Equipment Ltd. (UMEL), who had built the JIM in 1969 to the specific design laid out by Joseph Peress. His earlier atmospheric diving suit (ADS) *Tritonia* had been their main inspiration, and he had named the new ADS JIM after his chief diver, Jim Jarrett. Phil recalls, "Other diving companies like Comex had already been sniffing around UMEL. But the reason the JIM suit folks ended up selling to us was because all those other groups wanted to mothball the ADS so it wouldn't interfere with their saturation diving. Instead, I was saying, 'No, no, no. This is the future. Never mind a chicken in every pot, I want to see an ADS in every boat!' The UMEL guys were quite taken with that idea, plus they had run out of money. So we were able to buy all the rights to the JIM suit for Oceaneering.

After the successful demonstration of the *JIM* suit's ability to dive down to 280 m
(920 ft), connect control and emergency lines to the blow-out preventer (BOP), and its
emergency well and blow-out control capability, Oceaneering ramped up production to
meet demand. Jim English photo

We also acquired control of the development of the *WASP*, a mid-water ADS
from DHB Construction in Britain."

Beginning in the early 1970s, Phil took on the management of these
atmospheric diving systems for Oceaneering, principally through his Can-
Dive subsidiary. When it came time to put the *JIM* suit to work, one of the first
projects was with Panarctic Oil in 1976. It was on their HECLAM 125 well,
located in 915 ft / 280 m of sub-zero water off Melville Island in the Canadian
Arctic. The company was keen to get a permit to work there, but to do so they
had to demonstrate that they could service a blow-out preventer at great depth.
So, in the middle of winter, *JIM* was to dive down to the wellhead, connect
control and emergency lines to the BOP, and demonstrate emergency well and
blow-out control capability.

However, first the BOP had to be adapted so the *JIM* suit could access it
and make those connections. Then, it was all about schedules and logistics
details, Jim English recalls: "For safety, Can-Dive sent two full systems, about
three tons of suits, winches, umbilicals, hydraulic power units, communications, oxygen, and carbon dioxide absorbent, as well as other accessories.
Can-Dive also had to deploy a remote video camera to record the event in
order to prove Panarctic's capabilities. The variables were staggering, as no one

The star of this mid-winter Arctic show was the *JIM* suit; supporting cast was (left to right) Peter McKibbin, Doug Elsey, John Balch, Walt Thompson, Graham Hawkes, Tony Moore, Roland Schwartinsky, Mark Atherton and Jim English. Photo courtesy Jim English

knew how the magnesium-alloy ADS suit and its various components would react to -40° temperatures while in transit. As well, there were shipments of gear and people arriving from three countries. Then once the equipment got on the ice, it all had to get sorted and reassembled."

In all, Can-Dive made four successful dives in 1980. Walt Thompson completed two of the dives, spending 5 hours and 59 minutes on the bottom for the first dive. Peter McKibbin and Tony Moore also successfully completed a dive into Arctic waters. "At that time, the *JIM* had no thrusters or mid-water mobility," Jim explains. "But because we were working on ice, it was possible to just grab the 300-metre-long cable that connected the suit to the surface and move the diver around by hand to wherever he needed to go." Besides the successful mission, those working there vividly remember the cold, as well as encounters with Arctic fox and muskox when walking to work on the ice. "Men armed with weapons patrolled all the time on snowmobiles to keep polar bears away," Jim notes. "When the project successfully wrapped up, the bar bill for one night with nine thirsty divers and about half a dozen oil crew was $1,000."

As soon as Panarctic demonstrated successful ADS use, they jumped ahead of their competitors because they got drilling permits without endless red tape and paperwork. As well, once oil companies saw what the *JIM* suit could do, that it "only" cost about a million US dollars, that it could be easily moved around with a supply boat, *and* that it was cheaper to operate than using a saturation system, the orders came rolling in. Phil explains, "Everybody in the business was well aware that a six- to eight-hour shift dive to a thousand feet required an expensive saturation dive system." Phil ticks the costs off on his fingers:

- "Ten days of leasing a ship that can carry the 20–25 tons of a 'sat' system is half a million bucks. And the Sat system itself can easily cost $3–5 million US."
- "Rental and labour (the working diver, a standby diver and the bell tender) and everything else adds another $300,000 to $400,000, so the tally for six hours of work is now up to $800,000–900,000."
- "Then factor in the mixed gas since the diver has to be in that chamber for 10 days breathing a special oxy-helium mix. That much breathing gas alone costs something like $150,000–175,000."

Next, Phil excitedly details the alternative costs of using an atmospheric diving suit:

- "For starters, there's the cost of buying or renting a suit. Buying an ADS runs about one million dollars US, while a Sat system can easily cost $3–5 million. And an atmospheric diving suit requires almost nothing in terms of handling."
- "It takes about 15–20 minutes to get into the suit and get all your pre-dive checks done. There's no initial 24 hours of compression."
- "When the diver in the ADS is ready to submerge, it takes a little less than ten minutes for him to get down to a thousand feet and be ready to go!"
- "Then, the diver does the six hours of work, and it only takes ten minutes to come back up to the surface, with no decompression required."
- "And the expense of breathing regular gas is 25 bucks."

"What a deal! Companies figured that out pretty fast!"

Working through Oceaneering, Phil then hired UK engineer Graham Hawkes to work under him to develop a mid-water ADS called the WASP. Essentially it was a JIM suit with a single wide-leg tube instead of separate legs. "When Graham sent me his final calculations and drawings, he wanted to know when the prototype would be built." Phil explained that there wasn't going to be a prototype until Oceaneering had completed its backlog of JIM suit builds. The priority had to be filling a number of oilfield contracts Oceaneering had for the JIM suit.

That wasn't the answer Graham Hawkes wanted to hear, so he resigned. Then he hooked up with Offshore Services Electronics Limited (OSEL), the company that made winches for the JIM suits. The result was a WASP prototype, built on the sly. As soon as Oceaneering found out, the board called another emergency meeting to figure out what to do about it. Phil calmly stated, "I'm going sue their asses off because we own all title and rights to the WASP." As a result, he spent most of the next summer in high court in Britain, quite an experience in its own right.

"I felt really sorry for Graham because somehow he believed that he was the inventor, not just the engineer, of the WASP. In the end, Oceaneering won the case. But the way the British judicial system works, the judge also expected that we would come up with a compromise—a very different concept of 'winning' as we know it. That's how we wound up having OSEL build the WASP for Oceaneering exclusively, which saved them as a company and provided us with the WASPs we needed." A side story to all that legal wrangling was that Phil invited Graham to Hawaii as a standby diver during tests of the JIM suit with Sylvia Earle as pilot. And Graham and Sylvia ended up getting married!

The Oceaneering minnow swallows the whale

In the spring of 1971, the fledgling Oceaneering and its leadership faced a most unusual dilemma. The president of Divcon International called Lad Handelman with an opportunity—buying a company that was roughly five times the size of Oceaneering.

It was a complex scenario since Divcon was already in the midst of finalizing the sale of itself to the large French firm Comex. Indeed, the papers were ready for signing. But Divcon had second thoughts. Lad quickly realized that if the deal went through, the Comex-Divcon combination would lock up 80 per cent of the world diving market. Oceaneering would be left with the crumbs. Although the newly minted Oceaneering lacked the resources, organizational capacity and expertise to buy and run Divcon, there was no choice but to go ahead with the last-minute deal. Despite Divcon being a much larger concern on paper, Oceaneering was able to reach agreement to buy their equipment without buying the company itself, which was wound down shortly after when its president retired. The minnow had swallowed the whale.

The year that followed the acquisition was fraught with all manner of problems—covert sub-companies, equipment failures and operational mishaps. But by the end of that first year, Oceaneering had somehow managed to steer the hugely expanded company into reasonably calm waters, both financially and operationally. Equally important, Oceaneering was now one of the two largest diving companies in the world. Comex was their competition.

Lad clearly knew the company's strength: "Oceaneering consisted of hard-core, hard-working people who knew the business, with a management that understood the need to let the field people run the operation ... The most important guys in Oceaneering's world were those diving superintendents, not the guys back at corporate headquarters."

In 1974, Oceaneering, Perry Oceanographics and General Electric developed a series of two-man diving bells, rated for a depth of 3,000 ft (915 m) and designed for offshore oil-related work. Officially dubbed Atmospheric Roving Manipulator Systems or ARMS, they were essentially diving bells with thrusters. But the key feature was the GE force-feedback, spatially compliant manipulator arm. These cost a quarter of a million dollars each, but Phil states that they were worth it. Over the next several years, Perry built two ARMS bells to Oceaneering specs, Can-Dive built and delivered another, and an unfinished one is still in Can-Dive's bone yard.

By 1975, only six years after incorporation, Oceaneering's joint revenues had mushroomed from $600,000 to $55 million, and the company was operating in 24 international locations. That same year, the company went public. Looking at the larger industry picture, it was also when oilfield diving went from being primarily a labour-intensive business to a capital-intensive one.

That meant the balance slowly shifted from divers in the water to expanded support services, including machinery and often a surface ship. This change would become even more pronounced by the 1980s with the industry-wide adoption of ROVs and their associated technology.

But like any other marine venture, oil-patch work is feast or famine—or as Phil put it, "It's either chicken or feathers." And in late 1978, with the developing crisis in Iran, the industry faltered severely. Drill ships were docked, construction budgets slashed, and the service companies contracting with the oil and gas industry took a beating. Oceaneering was no exception. The company's income declined, as did its profits. To complicate matters, Lad Handelman also suffered a devastating skiing accident that left him a quadriplegic. While Oceaneering's management struggled to survive, the giant Chicago Bridge and Iron made a series of big stock purchases and attempted a hostile takeover of the company. Lad Handelman was able to kill the attempt, but in the process, he alienated some important board members.

Lad suddenly found himself out of the company as Mike Hughes took over control of the board. As a result, Lad dumped all of his Oceaneering stock in one day. He was determined to go his own way and eventually joined others who had also quit the company. They formed a new dive company, naming it Cal Dive after his original venture. Even as a quadriplegic, Lad worked with the new company as it began to grow.

Can-Dive logistics

By the mid-1970s, Can-Dive, operating as a subsidiary of Oceaneering, had nearly a decade of work experience to draw on, a string of contract successes, and office headquarters in North Vancouver. The company's early digs were functional, if unimpressive. Located on the upper floor of an older building, access into the building's four bays was via the back alley. The main bay housed mesh lockers for diving gear, a machine shop and an assembly, welding and storage area. A second bay had office space, a purchasing area and tool and dive gear storage. Another bay was a hodgepodge of office cubicles. Phil's office was in the front half of the fourth bay, with the back half for research and development, as well as oceanographic survey and electrical work. Even the back alley was an essential part of "the office." There was some parking, but primarily it was the locale for fabrication, storage and testing of all sorts of gear, from decompression chambers and diving bells to test tanks, handling systems, winches and any number of other gadgets and supplies.

Helmut Lanziner came to work for Can-Dive in 1970, bringing his sonar and electronic background. Phil set him up as a separate branch of the company. Helmut states, "Phil called it Can-Dive's Oceanographic Survey Division, but we had nothing to do with oceanography. I guess the right name should have been 'Technical Services' because in the beginning we were usually searching for and finding objects lost at sea." Helmut also recalls dives in the small submersible *Sea Otter*, often for cable inspections in BC or in the US.

"Phil said I could operate this new division almost like a separate company, so I got to hire my own people and I got a door that locked!" Helmut had three "regulars" among others on staff: Frank Tremblay, Roland Schwartinsky and Garry Kozak. Often Mark Atherton joined in, on loan from the Can-Dive side of things, and sometimes Willy Wilhelmsen came by to lend a hand with sonar requirements.

Can-Dive hired Jim English as its first full-time engineer in 1975. He explains: "From the start, the job was hands-on. Sometimes it included going out on an inspection team or a dive team if an engineering assessment was required. I also did a lot of bidding and estimating, as well as searching out needed equipment, materials or supplies for larger projects. And I put my engineering training to work building gadgets or tools or sometimes systems as needed. This was all before computers and word processors, so everything was hand-written, -drawn or -calculated. If it was a proposal or had to go on record, our secretaries typed it all up. Most of what we did was face to face. Major deals were done with handshakes and simple one-page letters."

Can-Dive also hired Doug Elsey as the company's resident geek; working in a Can-Dive office in Mississauga, he was an engineer who could also handle operations management. Jim recalls: "He recognized early on that computers would be good things and dazzled us with output from a Tandy machine and a Commodore 64."

Support work in the oilfield

Given that subsea work is highly cyclical, it becomes imperative for any savvy company to pay attention to emerging trends—and beginning in the 1960s and '70s, that meant oil and gas work. There was big money to be made supporting offshore drilling operations, essentially supplying divers and equipment to drill sites. As a result, Can-Dive's focus expanded dramatically from underwater construction to oil-rig support work. Generally, that meant ever-increasing depths, tight timelines and unbelievable extremes in weather.

The quest for oil in Canadian and Arctic waters came somewhat later, aided by Canadian government Arctic drilling incentives that started in 1974. Big oil companies such as Panarctic Oil, Esso and Dome Petroleum brought in fleets of drilling rigs, equipment and drill ships—and prosperity to nearby regions. From 1976 to 1982, Panarctic drilled 38 offshore wells from the ice platforms they created, at a cost of $22–23 million per well. This economic frenzy lasted for a decade, until oil prices dropped and the Canadian government ended drilling incentives.

From 1977–78, Panarctic also contracted with Can-Dive to provide extensive diving support for the world's first under-ice gas pipeline installation—a project known as the Drake F-76 pipeline. Once all the equipment was delivered onto the ice, Jim English had to make sure it all worked. With the temperature at -30° C, he recalls that the first few weeks were spent turning divers into popsicles! Deeper diving work was actually warmer for them, but more hazardous. Fortunately, the Drake F-76 pipeline proved a major success.

Based on that accomplishment, Can-Dive secured a contract with Polar Gas, a consortium of oil companies. One of the project's benefits was that Can-Dive designed and built the *Constructor*, a hybrid one-atmosphere/saturation diving bell fitted with thrusters and robotic manipulators. It would see conversion and service in a number of subsequent projects.

By the mid-1970s, Can-Dive had become the main diving contractor supporting oil and gas operations on the East Coast, operating not only bell diving systems on various drilling rigs, but also offering bounce diving (repeated

dives separated by short intervals to depths or no more than 20 m/65 ft).
Based largely on their reputation and experience in cold water, the company
won contracts with Dome Petroleum and their drilling contractor, Canadian
Marine Drilling (CANMAR), in the Beaufort Sea. Most drilling work there
happened from June to late September. Maintenance and repair took place
during the winter when the drill ships and all the support vessels were frozen
in the ice. The CANMAR operation spun off into other Arctic work with Esso
and Gulf Canada.

"Unless you've been there, it's hard to imagine the advance planning that
had to go on for these jobs or the thousands of dives done in very harsh conditions," Jim English explains. "Divers were working by feel in zero visibility with
wellhead and blow-out preventer systems often buried below the mud line. If
a diver lost a wrench, he had to find it or come back to the diving bell and wait
for a replacement to be lowered, which took a lot of time. A rough estimate
of what it cost a company to drill in the short summer season was upwards of
$1,000 per minute, 24/7. It was outrageous and generated immense tension for
all workers. If a tool pusher was behind schedule in drilling a hole, and a diver
down below lost a wrench, he loved blaming the diver for the entire problem.
The notation in his log was WOD (Waiting On Diver).

"If you had equipment problems for any length of time, the cost was huge.
For example, on the surface there are big pieces of subsea blow-out preventer
equipment with 15 or 20 guys putting all the hoses on each piece, getting it
ready. Then they deploy it and start drilling. Next, something breaks. So now
instead of 15 or 20 guys on dry land doing the work, they send two guys down
in a bell and one of them goes out to find and fix the problem in thick muck
and zero visibility. Of course, Topside yells at him if it takes too long. So the
time pressures and the money pressures were huge. That's what was getting
people killed back in the early days."

Jim adds: "The pressure to get the job done regardless of the cost—financially or in terms of safety—is one of the big things that drove the industry.
If somebody phoned you up at midnight and said, 'We want 10 people here
tomorrow morning at 6 o'clock and we have to have this job done by noon,'
you would do anything to meet that deadline. If you got the job done on time,
you'd get a pat on the back and a big paycheque. If you failed, it was very bad

news. Folks were on call 24/7, and that kind of pressure was daily life for all the managers and divers."

Can-Dive's new product: *Deep Rover*

Over the years, various projects, finances and decisions see-sawed back and forth between Oceaneering as the parent company and its subsidiaries, in this case Can-Dive. In addition to contractual work, either in subsea construction or oil and gas support work, Phil Nuytten also researched and pushed the development and utilization of new products. In 1982 Can-Dive went ahead with building its one-person submersible *Deep Rover*, with a depth rating of 3,000 ft (915 m).

Phil did up drawings for *Deep Rover* and then hired Graham Hawkes to work on it as engineer. Phil had several goals for this submersible, beginning with an acrylic sphere that would provide outstanding visibility in all directions. He also wanted a cleverly designed frame of high-density polypropylene, and thrusters that would deliver easy steering, allowing the sub to "fly." What it lacked in depth rating, it would more than make up for in manoeuvrability. And its reasonable price tag was perfect for scientific institutes or oil and gas firms with less than deep pockets.

Can-Dive put *Deep Rover* to work on a number of contracts, with Steve Fuzessery becoming chief pilot and technician. The sub's first major job was a year-long opportunity, with Dr. Bruce Robison conducting the first biological exploration of the waters of Monterey Bay. One immediate result was that engineer Derek Baylis designed and constructed an underwater housing for a broadcast-quality camera, which would allow Robison to capture images of the Bay's watery wonders. David Packard (one of the founders of Hewlett Packard) was a benefactor of the nearby Monterey Bay Aquarium, and the bigger impact of Robison's work was in broadening Packard's thinking to include a research program with a much broader agenda. The result was the Monterey Bay Aquarium Research Institute (MBARI). Best of all was seeing the excitement of scientists as they emerged from a dive in *Deep Rover* with original data, samples and tales of new species.

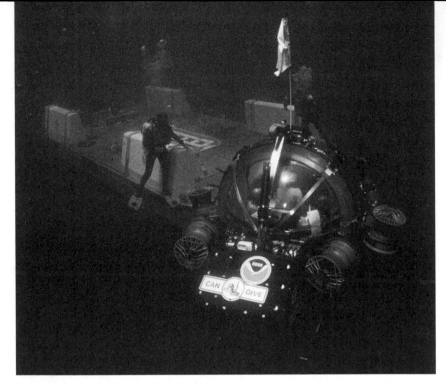

Deep Rover taking off from the Hawaii Underwater Research Lab's submerged launch and recovery pad. Graham Hawkes, Sylvia Earle and Phil Nuytten would each take *Deep Rover* to 3,000 ft (915 m) on the submersible's first full-depth open-water test dives. Personal collection of Jim English

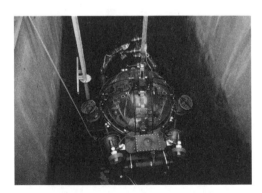

Deep Rover was tasked with inspecting the deep, dark, long tunnels at Niagara Falls, on both the New York and Ontario sides of the Niagara River. It was not a job for the faint of heart or the claustrophobic. Personal collection of Jim English

From 1988–89 *Deep Rover* was integral to a scientific quest to discover evidence of hydrothermal venting in Oregon's pristine Crater Lake. Despite limited physical access and a tight time frame during the two-year project, *Deep Rover* was able to pinpoint the hydrothermal vents in the lake with high temperature and salinity gradients, bacterial mats, and brine pools. Doing so helped quash a planned geothermal energy development nearby.

The tunnel team, along with *Aquarius*, a refitted *Sea Otter*, and *Deep Rover*. Not shown are two small *Phantom* ROVs. Personal collection of Jim English

An ambitious non-science job was *Deep Rover*'s horizontal work in Niagara Falls' immense and long diversion tunnels. Starting in 1987, on both the Canadian and US side of the falls, water gets diverted at night from the Niagara River above the falls into massive underground tunnels that lead to reservoirs. The purpose is twofold. Reducing the water flow over the falls at night helps control the rate of erosion at the lip of the falls. But the reservoirs also allows hydroelectric power systems to release the water in a controlled flow during the daylight periods of peak power demand. With Steve Fuzessery at the controls, *Deep Rover* safely inspected the interior of those massive tunnels. A small *Phantom 300* ROV was added, allowing the pilot to inspect side shafts from the *Deep Rover*. The success of this tunnel scrutiny led to more work for the submersible.

An exit from Oceaneering

Oceaneering International was doing well, and by 1984 would post sales of $1 billion, with 5,000 employees in 35 countries. Despite this rosy financial picture, Phil Nuytten was aware that his vision of the future differed significantly from that of his partners in the parent company. He wanted to concentrate on ADS and submersibles. Oceaneering wanted to focus on the lucrative saturation-diving business and ROV operations. There were also issues with the new senior

The ADS *Newtsuit* was a long-time dream for Phil Nuytten. He's shown here getting a final communication check before descending into Kyuquot Sound. Personal collection of Phil Nuytten

management's direction. And, contrary to agreement, Solus Ocean Systems, an ROV company that Oceaneering had acquired, and Montreal's SNC-Lavalin, were all of a sudden competing with Can-Dive for work in eastern Canada.

Phil also wanted more time—to be with his family and to tinker, invent and solve underwater engineering puzzles. So in 1984 Phil sold his Oceaneering shares and used those millions to buy back Can-Dive from Oceaneering, repatriating the company to Canada. However, 1984 also marked a downturn in oil prices, withdrawal of government drilling incentives, and a corresponding loss of revenue from lucrative Arctic operations. It was a tough time for Can-Dive to jump out of the company boat.

After exiting from Oceaneering, Phil immediately began work on a new ADS, one with better, more flexible joints that would enable the diver inside to do delicate work despite the intense underwater squeeze at depth. Phil's new rotary joints were patented in the US and Canada in 1982 and 1985. That same year, Mike Humphrey moved to Canada to work for Phil on what would become the first truly dexterous ADS. It was dubbed the *Newtsuit*, a catchy name based on the pronunciation of Phil Nuytten's last name and also a tongue-in-cheek recognition of the aquatic salamander, or newt.

By 1986, the final prototype of the *Newtsuit* was completed in Dartmouth, NS, and shipped to Vancouver, enabling both *Deep Rover* and the *Newtsuit* to fascinate thousands who came, touched and gawked at these underwater showpieces during EXPO '86. That same year, Phil set up International Hard Suits Inc. as a subsidiary of Can-Dive. The *Newtsuit* went into full production in 1988.

Complex construction work

As Can-Dive grew and diversified, it took on bigger contracts. The repatriated company secured local construction contracts in addition to its oil and gas work. The most challenging of those jobs was for the Vancouver Island natural gas pipeline. Running from 1989 to 1995, it was a lengthy, complex job. And although Can-Dive completed it successfully, it nearly broke the company. Westcoast Energy and Alberta Energy had secured permits to install a natural gas pipeline running some 590 km (366 miles) from the mainland to Vancouver Island. The project involved two major water crossings (on either side of Texada Island, mid-way across the Strait of Georgia), and Can-Dive was to survey both route crossings, utilizing divers, an ROV and *Deep Rover*.

The steel pipeline itself came off the stern of the reel ship *Apache*. It was the first self-propelled, dynamically positioned vessel designed for laying steel pipeline from a reel. One section of pipe would be laid, then welded to the next section, and the process repeated until the crossings were complete. However, irregularities in seabed terrain meant that some sections of pipe were suspended above the bottom unsupported. That created stress on the pipe at these points, so some sort of span support was needed. Can-Dive bid for and won a second contract to provide this.

The challenge of this new contract was the variety of span depths—some as deep as 450 m (1,475 ft) and some as little as 4 m (13 ft). The solution was to install supports consisting of fabric bags pumped full of cement grout. But nothing about the grout bags was simple. Filling, placing and securing those bags involved a large ocean-going barge fitted out with a deepwater mooring system, a crane, an air diving system, a mixed-gas diving system and a dual ADS system. Later on, a full ROV system was added as well. And on board the barge were silos of dry grout, a pump and a mixing plant, plus all the auxiliary gear. Finally there were permits, vessel charters and insurance.

Besides the underwater supports, Can-Dive also had to install stabilization mats, essentially large "mattresses" laid on top of the exposed sections of pipe to provide protection. Each one weighed 2,500 kg (5,500 lbs) and had to be carefully positioned using sonar. Mark Atherton was put in charge of that. "The problem was the mat had to land just so on each section of pipe," he explains. "It had to overlap the pipe sufficiently on all sides but couldn't cover any of the pipe's anodes that prevented corrosion. The other trick was to get the mat aligned over the pipe before setting it down into position." Mark finally figured out placement logistics and how sonar could assist. "So Can-Dive started fabricating the mattresses, and we began laying mats faster than anyone thought possible." Mark laughs wryly. "It was all working perfectly … until the barge sank!"

As every contractor knows, no construction job ever goes smoothly. But this job was a nightmare. First, Can-Dive's barge overturned and sank in the midst of a major winter storm. Everything—crane, equipment and gear—was lost. Fortunately, the pipeline wasn't damaged, so mat work could continue. On other pipeline sections Can-Dive's ROV was working 24/7. Then one night the ROV's umbilical got hopelessly entangled in the support ship's propellers. A Can-Dive crew spent the next 36 hours straight unsnarling that mess. Problems continued with diving suits and then with the grout.

At its worst, pipeline work was shut down about 50 per cent of the time, at a cost of tens of thousands of dollars per day. Can-Dive discussed the problems and the resultant delays with the contractual parent company and were told the extra work would be covered. Alas, they failed to get that commitment in writing. And even though Can-Dive installed 149 of the protective mattresses and successfully completed the job, the lack of a signed agreement regarding compensation for the extra work proved to be a significant snag. When the parent company, Morrison-Knudsen/Northern Construction, refused to pay, the subsequent legal case dragged on for three years. Can-Dive won in court, but then their opponents appealed and the decision was reversed on a technicality. The upshot was a significant financial loss for Can-Dive.

Going public: International Hard Suits becomes Hard Suits Inc.

With finances extremely sketchy in 1990–91, Phil opted to close Can-Dive rather than have to declare bankruptcy. He created a new version of the company and bought the assets of the old company. Then, in early 1992, driven by the continued need for increased financial support to develop the *Newtsuit* product line, the Can-Dive Services and International Hard Suits companies were combined into a publicly traded company on the Vancouver Stock Exchange with the new name of Hard Suits Inc. Strangely enough, it meant that Can-Dive was now a subsidiary of Hard Suits Inc., although the marine contracting business carried on as usual. Hard Suits Inc. continued with ADS manufacturing and sales, but the new company also delved into engineering and new technology development.

Unfortunately, after Hard Suits Inc. went public, Phil's life became even more complicated. He struggled to support the suit and other developments based on revenue from operations, as well as deal with the deluge of paperwork that a publicly traded company required. Setbacks in ADS operations and a shutdown in Arctic oil exploration strangled cash flow. The lawsuit over the pipeline-cost overruns was another nail in the coffin as the company's overall financial health became critical. With a lot of juggling, Phil somehow managed to keep everything afloat.

In August 1992 Phil launched Nuytco Research, a separate Canadian company that he hoped would allow him to spend more time doing research rather than simply building and selling atmospheric diving suits and vehicles. In 1994 Scott Lyons became president of Hard Suits Inc., but only briefly. A year later, Phil hired Rod Stanley, whom he knew as a former employee at Oceaneering. He hoped Rod would handle the mountain of paperwork required for a publicly traded company and find new funding or partnerships that would allow the company to expand or provide capital for existing projects.

By early 1996, the plot thickened. Rod and Phil were in negotiations to sell Hard Suits Inc. The prospective buyer was Cal Dive International, the very company that had formed after Lad Handelman's split with Oceaneering. Like Phil he had not only gone back to the original name but had again become a major player. The papers were on the table ready for signing. But at the last minute Phil learned more about backroom misinformation relating to the sale, and the deal collapsed. Rod Stanley resigned, and his long-time relationship with Phil evaporated.

Things got worse. Shortly after the Cal Dive deal collapsed, Hard Suits Inc. with all related assets, intellectual property and contracts became the subject of a hostile takeover by American Oilfield Divers, headquartered in Louisiana. They were particularly eager to acquire Hard Suit's ADS, submarine rescue connections and engineering technology, much of which would enhance their position in the Gulf of Mexico and deepwater offshore market.

American Oilfield Divers saw the company's management upheaval, shareholder dissatisfaction and general cash bind as a golden opportunity to acquire Hard Suits Inc. on the cheap. And that's exactly what they did, buying up the majority of shares at above the listed market price. Ultimately, Hard Suits was bought for less than a tenth of the negotiating price that had been offered by Cal Dive. After initiating and completing the takeover, the company subsequently hired Rod Stanley to manage the transition for them. Rod took over as the CEO of American Oilfield Divers in December 1996.

Phil was blindsided by the hostile takeover—and Rod's involvement in particular. "That was a real tough one," he acknowledges. As a result, Phil wanted nothing to do with the American Oilfield Divers/Hard Suits organization. He took his buyout money and simply carried on operating Nuytco Research. He had managed to extricate Can-Dive from Hard Suits Inc. before the takeover, so retained the rights to that name, which continues to operate to this day. Phil also kept his ADS patents and the *Deep Rover*, *Sea Otter* and *Aquarius* submersibles. He refused to sign a non-compete clause, determined to continue his own atmospheric diving suit work as well as designing small submersibles.

Perhaps Phil Nuytten's ability to shift his focus, a trait he'd already recognized as a kid, helped him survive Can-Dive's rollercoaster financing and the see-saw of contracts, which ranged from underwater construction to servicing oil wells. Fortunately, there has been another side of Can-Dive's diverse work: the film business. That has involved providing diver support, highly specialized equipment, and occasionally even doing the filming for TV shows, documentaries and ads. But feature film work is always the craziest. On the outside, movie production is tinged with Hollywood tinsel and glamour.

Behind the scenes, the work is incredibly demanding and repetitive, especially for scenes shot underwater.

In the late 1980s, movie director James Cameron started work on his epic underwater film *The Abyss*. Cameron asked Phil to come to Fox Studios to talk about commercial diving. Also there was Al Giddings, a long-time friend of Phil's, who was slated to shoot the film. Quietly, Phil asked Al, "What's with this guy?" Al replied, "I don't know, but let's give him a run for his money." Cameron explained that he wanted to know what commercial diving talk was like in order to use authentic-sounding jargon in the film. Opening his notebook, he asked, "So how do you describe a dangerous situation?" Phil winked at Al and then answered seriously, "We give it a sphincter factor. If it's a really bad one, that would be a sphincter factor of ten." Phil recalls, "Al was going crazy trying not to laugh, but if you watch *The Abyss*, our sphincter factor is in the movie."

Can-Dive was also tasked with fabricating two futuristic vehicles. "The first was *Flat Bed*, the big lockout work sub in the movie. Inside *Flat Bed* we put a 100-horsepower ROV that moved it around. And the front end came from a damaged front dome of *Deep Rover*. We also created the *Cab* subs in the movie out of *Aquarius*. Don Pennington flew up from LA and took pictures and measurements of *Aquarius* and then built a full-sized fibreglass replica that became the *Cab* sub duplicate. In the movie there are three of those subs—*Cab 1* was white, *Cab 2* was yellow, and *Cab 3* was orange. Some people are convinced they could tell them apart, but in fact there was really only one sub. The props department would just paint it different colours between scenes!"

Phil details other scenes of movie subterfuge, particularly when lead actress Mary Elizabeth Mastrantonio is "driving" *Deep Rover*. "Her controls were fake. One of our guys in the back of the sub was hooked up in communication with her and saying things like, 'OK, start to move the joystick to the left, to the left, to the left. Now, back towards you, back.' When you look at that movie, you think she's piloting the submersible. Even I can't tell she's not flying it!"

Filming and dive operations required amazing human choreography that translated into a brutal work schedule. "There was a base crew of 30 to 50 divers, with safety divers for each actor, plus our six-person team. The surface support teams were almost uncountable, with upwards of 500 people on site every day,"

Jim English recalls. "It often took days to shoot a scene that might last less than a minute on screen. Hundred-hour, seven-day work weeks became the norm. It is a credit to the Can-Dive crew that they never once delayed the shooting, even when Cameron changed his mind at a moment's notice."

Sometime after *The Abyss*, and later Cameron's *Titanic*, won Oscars, Phil recalls being seated together with Cameron at the annual Explorer's Club dinner and awards ceremony. Phil was about to receive the Lowell Thomas Medal for ocean exploration, and Cameron was slated to receive an award as well. "Jim leaned over and asked me, 'Are you nervous?' I said, 'Why would I be nervous? I know everybody here.' And he told me, 'Well, I think I'm a bit nervous.' I broke out laughing, and he asked, 'What's funny about that?' And I said, 'Jeez, Jim, are you kidding? It wasn't that long ago that you won all those Academy Awards in front of swillions of people watching on TV, and you sure as hell didn't look nervous then!' He looked at me and said, 'That's all bullshit. This is real.'"

Memorable projects: *REMORA*, *DeepWorker*, the *Newtsuit* and *Exosuit*

Phil Nuytten is a survivor. He can boast six decades of underwater work and more than five decades of corporate history. During that time, several projects stand out in Phil's mind as exemplary. One is REMORA, a submarine rescue system built for the Royal Australian Navy on an incredibly tight deadline. Its formal name was REMotely Operated Rescue Appliance, but a remora is also a small sucker fish that attaches itself to sharks, an apt tag for a rescue system designed to attach to a larger sub in distress.

The Australian Navy already had a firm December 1995 launch date for its first diesel-electric submarine, but then decided to add a rescue system. The catch was they needed it delivered by the same launch date. Commander Frank Owen and Dusty Miller took on the challenge. In January 1995 Dusty Miller approached Can-Dive and discussed possible ideas as well as the project's very tight deadline. With only one day to bash around ideas, they came up with the world's first remotely operated rescue vehicle (RORV), with all functions running from the surface via an umbilical. Two attendants would handle life support, hatch ops and the transfer of rescued personnel. The hull would be re-worked from the *Constructor* diving bell that Can-Dive had acquired after a previous oilfield contract. And the vehicle would use an articulated mating skirt to attach to a downed sub. The skirt, plus a variant of Phil's *Newtsuit* rotary joint, would guarantee that the rescue bell could mate with a sub sitting at any angle from 0 to 60 degrees.

The remotely operated rescue vehicle *REMORA* features an articulated mating skirt, which can attach to a downed sub.
Personal collection of Phil Nuytten and Jim English

With only days to detail and submit a concept, cost outline and hoped-for schedule, the final bid was $12 million. Jim English recalls that the Rough Order of Magnitude (ROM) pricing approach was, in reality, a WAG (Wild-Assed Guess). What no one doubted was that getting the contract would be a significant shot in the arm for Can-Dive. While waiting to hear the results of the bid, the company forged ahead with setting up a team of both employees and outside hires. International Submarine Engineering, with James McFarlane Sr., was the major external player.

In June, Can-Dive got the two-page approval to start work—and a deadline for completion of November 24, at which point it would have to be ready to load onto trucks to be driven to Los Angeles, the nearest point from which it could be flown on a large equipment transport to Australia. That meant only 22 weeks to deliver, fully certify, test and be ready to ship the world's first remotely operated submarine rescue vehicle. Phil's idea of using an articulated skirt to mate with a downed sub was a new, complex concept. But the final product worked so well that Phil received a patent for it in 1998. Employees regularly worked a 60- to 70-hour week, with only Sundays off. Jim English recalls, "One thing we did that paid off was to have a Beer and Band-Aid session at the shop every Friday afternoon. A lot got discussed and sorted out during those sessions."

Finally, it was time for water trials in Indian Arm, followed by deepwater success in Jervis Inlet. With only four days remaining, all the bits and pieces of REMORA were cleaned, inventoried and loaded onto trucks heading to Los Angeles. The last truck left North Vancouver on November 24, 1995, at around 6 p.m. Jim recalls, "We made the deadline. We pulled off the impossible."

Phil's next memorable project was the one-person submersible *DeepWorker*, developed by Nuytco Research. Sixteen years earlier, Can-Dive had built *Deep Rover*, a submersible that Dr. Sylvia Earle had often piloted. She soon became intimately acquainted with its successor, launched in 1998. It featured an on-board computer and convenient touch-screen controls so the pilot could concentrate on work. Forward/backward, up/down motions were easily controlled by foot pedals, leaving hands and arms free to work manipulators or camera equipment. Four thrusters gave *DeepWorker* amazing manoeuvrability. And there were 2,000 or 3,000-ft (610/915 m) options for depth. Another advantage was that the submersible was extremely lightweight so could be transported by trailer and launched from a wide variety of ship platforms. *Dual DeepWorker* followed soon after, allowing two scientists to work side-by-side.

DeepWorker's convenient design and portability helped the submersible gain an enviable global reputation. Personal collection of Phil Nuytten

The new submersible quickly gained a global reputation, helped by its contribution to the success of the Sustainable Seas Expedition. This five-year initiative by the National Geographic Society and NOAA studied deep ocean environmental impact. Beginning in 1999, *DeepWorker* allowed notable scientists like Dr. Earle to descend deeper (2,000 ft/610 m) and remain submerged longer (up to 12 hours) while exploring and monitoring US National Marine Sanctuaries. Recently, NASA and the Canadian Space Agency have been using *DeepWorker* to explore early forms of life, known as microbialites, found in Pavilion and Kelly Lakes in the BC Interior.

―――――――――

Name one product that identifies Phil Nuytten and it's likely his invention of the *Newtsuit*, which went into full production in 1988. Can-Dive's Bruce Fuoco became lead pilot and champion for the *Newstuit*, demonstrating its capabilities and inventing tools to make jobs easier and faster. He trained hundreds of pilots as Can-Dive sold suits to navies and companies in France, Italy, Japan, Norway, Turkey, Russia, Greece, Canada and the US. Phil notes proudly: "I think we've trained nine Canadian astronauts for the Canadian Space Agency, (CSA) and a couple from NASA in these suits, using them as a gravity simulator." Julie Payette was one such trainee. In addition to her subsequent space flights and appointment as chief astronaut for the Canadian Space Agency, and then as Governor General of Canada, Bruce remembers that she was also one of the best *Newtsuit* pilots he ever trained.

Phil is always keen to explain that the beauty of an ADS—and its advantage over any type of surface-supplied or saturation diving—is that you can go up and down to great depth any time you need, with no requirement for pre-breathing oxygen or decompression. "Travelling 365 metres from ocean floor to surface takes only about 10 minutes instead of 10 days. And when you get to the surface, you can immediately carry on with a normal life, with no need to be locked in a can while decompressing." Given Phil's diving history, that's an advantage he knows well and certainly appreciates.

Jim English recalls one "unusual" *Newtsuit* job in Greece, where the year-long mission was to help install a 3-m (10-ft) diameter concrete sewer outfall, a line that serviced the entire Athens region. Working conditions on the barge were filthy, and safety was clearly not a priority of the local contractor. In fact, three workers were killed on that job. Jim states: "We fought for and controlled our own destiny and completed the job unscathed. But there were certainly other challenges. Each day in the late afternoon, it seemed that everyone in Greece flushed their toilets at the same time. The existing sewer line, which ended upstream from our barge, dumped all sorts of unmentionables around us. It was definitely not the vision of diving in the Mediterranean that

Phil Nuytten is known for his creative design, whether it's fabricating a prototype *Abyss* dive mask to provide greater visibility of the actor for the film industry or making adaptions that created his next ADS upgrade, the *Exosuit*. Photo by Vickie Jensen

I'd gotten from watching Jacques Cousteau films."

While Phil Nuytten is certainly grateful for his life, "contentment" is not a word to associate with his career aspirations. So with the *Newtsuit* already hard at work, Phil began dreaming of an updated version. He finished a prototype in 2010, and the *Exosuit* was unveiled at a New Orleans trade show in 2012. Both the *Exosuit* and *Newtsuit* have aluminum-alloy bodies, a 1,000-ft (305-m) depth rating and the dexterity to perform delicate work. Both ADS suits also feature interchangeable hand pods, so it's easy to switch from manipulators for heavy-duty work to force-feedback "hands" for gentle jobs.

But the *Exosuit* boasts significant improvements that include upgraded electronics, far more powerful thrusters, advanced rotary joints, and a teardrop shaped visor that positions the arms closer together for more comfortable extended work times. Phil describes the *Exosuit* as "part diving suit, part swim suit, part space suit." Never one to let his imagination rest, Phil is also finishing up *Ironsuit 2000*. "It's a beast!" he says of the latest ADS suit. Constructed of high-strength iron, it's designed to allow work down to 2,000 ft (610 m), especially as a military submarine rescue asset.

Phil is also proud of the amazing cast of skilled characters who form the core of Can-Dive and Nuytco Research today. Jeff Heaton is Phil's second-in-command, the youngest son of his long-time business partner Don Leo Jonathan;

Dave Porter handles field projects; Mark Arnott is the head of electronics, working with Jeff Rozen; Mo Kizemi is in charge of the machine shop; and Bill Liddell is the company's accountant. Phil's daughter Virginia Cowell is his personal assistant at Nuytco and the editor of *Diver* magazine, the longest-established sport scuba-dive magazine in North America, which Phil also owns and publishes.

Can-Dive construction work provides a relatively stable income. Nuytco handles the manned submersible and ADS markets, including hardware, sales and operations of dive suits, as well as other smaller high-technology product lines. During one of many interviews, Phil takes me through the company's fabrication area. There are four *DeepWorker* subs in various stages of assembly. "We produce a lot of our thrusters now, doing a batch of fifty at a time. They're unbelievably powerful. And we build the lithium battery packs for them that deliver three times the amount of power we used to get out of lead acid batteries at only one third the weight. That's great for these small *Workers*. These are 3,000-foot-depth subs, built out of tough HY80 steel to handle the extra depth. Our standard *DeepWorker* subs are rated for 2,000 ft. They look exactly the same as the other ones, so we have to put tape on them to tell them apart."

Next, Phil points out a compact ROV: "That's the most powerful electric ROV we've ever built—and it's all done online, all on computer, and then cut out by that machine over there. So there's no need for a factory—we're it!" Originally dubbed the *NewtROV*, the vehicle is now being reconfigured as the *NewtROV Leveller*, intended for global navies with small submarine fleets that can't afford the high cost of a system like the US Navy's PRMS rescue vehicle, a very large, very expensive version of the original REMORA. "The *Leveller* even has a REMORA-like articulated skirt that can latch onto the disabled sub; then the upper part extends to match with the rescue vehicle, providing a safe exit if required."

We pass by *Exosuits* in production and then detour into another section with totem poles in the process of being restored, along with masks and feast bowls being carved for Phil's next potlatch in Alert Bay. Then we move on to Phil's amazing collection of historic diving gear, each with a story, as well as vintage submersibles. As we walk, his commentary ranges from equipment to expertise, machinery and inventions to diving magazines. It's abundant proof that decades of hard work and risk, coupled with perseverance and imagination, ensure that boyhood dreams of adventure can come true.

In 2008 Dr. David McGee wrote a historical assessment entitled "Underwater Mobility in Canada: 1800 to 2007" for the Canada Science and Technology Museums Corporation (now Ingenium Canada). He concludes: "There was a greater expansion in underwater mobility in the years 1945 to 1980 than there was in all of previously recorded human history. A vast array of new technologies matured, ranging from scuba gear to saturation diving, and atmospheric suits to manned submersibles—to say nothing of the even vaster array of sensors, sonars, samplers, cameras, manipulator arms and other systems that were invented. Whereas only a handful of humans had previously ever been beneath the sea, hundreds of thousands now dove the oceans and hundreds of thousands more gained an awareness of the oceans through photographs and films." He concludes his report by saying that "Canadians like Phil Nuytten, Joe MacInnis and the leaders of International Hydrodynamics became leaders in a fabulous era that has not been equaled."

CHAPTER 6

T. Thompson Ltd.:
The importance of connections

1972–2018

There's always plenty of talk about risky underwater jobs once they finally get completed, or a difficult contract that somehow gets wrapped up, or even a new invention once it's on the market. But there's less talk about how sonar products got encouraged, how cables, cameras or couplers were sourced and delivered, or how contracts, especially major military ones that often dragged on for years, got facilitated. That was Terry Thompson's special skill, aided by his wife, Linda. She recalls the 1972 start of the company that they simply named T. Thompson Ltd.

"Terry and I immigrated to BC from the US in 1972. Terry was a US Navy SEAL, and we met five days after he got out of the Navy." After they married, Terry entered Purdue University, in Indiana. Upon graduation, he joined other Navy buddies working for Westinghouse on their *Deep Star* project, a research submersible rated to 4,000 ft (1,220 m). Terry handled marketing and purchasing for the project, so was involved with many companies while procuring underwater communication systems, lighting systems and underwater cameras.

In the 1960s, the Vietnam War started to siphon off research money, and then the US government shut down research programs. Westinghouse transferred the Thompsons back to Baltimore. Several ex-*Deep Star* people they knew, including Robert Bradley and Ron Church, had moved to Vancouver or worked there on HYCO's *Pisces* submersibles. Linda recalls, "Bob and Terry regularly talked on the telephone, and he kept telling Terry that he should move up to BC." So, a few months later, Terry gave his notice. The couple packed everything into a U-Haul and moved to Vancouver.

Terry Thompson shown with OceanWorks'
HS2000. Personal collection of Jim English

"Because Terry had been in the *Deep Star* project, he knew a lot of people and a lot of the manufacturers," Linda says. "When we immigrated, HYCO was still manufacturing submersibles and needed a lot of equipment, so Terry helped with that. We also facilitated products with companies that HYCO was doing business with, like Willy Wilhelmsen's sonar at Mesotech. And there was Phil Nuytten with Dave Porter and Helmut Lanziner working with him, too. And soon International Submarine Engineering got going."

Jim English gratefully recalls the work of T. Thompson Ltd. "They brokered acquisition of a wide variety of equipment, such as underwater cameras, sonars, umbilical and tether systems, lights and a huge number of other subsea products. Terry played a significant role getting work for Can-Dive, Hard Suits and OceanWorks; his company also served as the agent for sales of the *Newtsuit/Hardsuit* ADS and our submarine rescue systems to the US Navy, as well as selected overseas navies.

"Terry opened many doors for us in the international military marketplace, especially in the US," Jim adds. "He seemed to know everyone from sailors up to admirals. He knew how the Navy system worked. As a result, he was instrumental in landing OceanWorks's two biggest contracts with the US Navy. Terry worked and travelled tirelessly with me on sales calls, providing input for proposals and nurturing customer relations all over the world. In our 30 years of working together, I don't think I ever went more than a few weeks without talking to him about some customer or some technical or logistics issue. He always went beyond the call of duty."

Initially, Linda wasn't fully involved at the company because they had two young daughters, but she recalls that "many times, Terry and I worried there just wasn't enough money to continue; we would be praying a commission cheque would come in. We took risks all the time, some financial and some just leaps of faith. We were very fortunate and it worked out.

"We weren't company representatives—we were manufacturers' reps for all different types of equipment and companies. So we travelled the world doing that and going to trade shows all over. We usually had anywhere from 20 to 30 manufacturers, all of whom had very specialized equipment or products. So we had to know about those products but also had to understand which one might work best for a particular need. I would do all the research and background work on the equipment.

"Terry was often a year or two ahead of product development in his thinking. If there wasn't a product or piece of equipment already, then Terry would figure out how to get that need met. He'd say, 'What we need is this,' and he'd describe an instrument that would do something, and then start talking to different manufacturers like Mesotech. Some sonar systems came out of those discussions. Or sometimes, someone needed a product so he would talk to a group of people and that often resulted in a new product being developed. Terry got along with a lot of people, and he was very honest. He didn't know everything, but if he didn't know he'd say so. Then he'd add, 'Let me find out.' And he always followed through on the things he said he'd do."

T. Thompson Ltd. had a large territory, representing many products in all of Canada. For others, it was usually down the West Coast. Linda adds, "And, of course, Terry's Navy connections didn't hurt either. Terry was always travelling a lot, but if there was a trade show, I'd go along and rent a booth for us. Then we'd have a hospitality suite with about ten manufactures all pitching in to help with the cocktail expenses, and we'd invite everybody that went to the trade show. Everybody had a good time and met everybody, too. The reality was that our business depended on the people you met but also on knowing the manufacturers—and their products."

During the eight years that Doug Huntington was in British Columbia, he worked for HYCO, then ISE, and finally for T. Thompson Ltd., leaving only to return to Britain. "Terry and Linda included me and my family in their family circle, which was great, especially when I was travelling. They also put me

Linda Thompson in 2018, shown with her favourite silver bracelet. "Phil Nuytten carved that for me in 1978. I wear it almost every day." Photo by Vickie Jensen

forward to be president of the BC section of the Marine Technology Society. That's another way I got to know their client base in marine technology and ocean science. T. Thompson represented various clients throughout the Hawaiian Islands, Alberta, the Pacific Northwest and Alaska. So for a couple of years I was their sales guy on the road. Bill Gronvold of Seattle handled their US sales."

For over four decades, T. Thompson Ltd. played an integral role in facilitating the contracts, inventions and sales of major BC subsea companies, a list that included HYCO, Can-Dive, International Submarine Engineering, OceanWorks, Mesotech and Imagenex. If the business was exciting, it was also one that Terry and Linda lived 24 hours a day.

Terry died in 2005. As Linda looks at the new technology today, she thinks, "Oh, Terry would have really enjoyed all the things that have happened." In 2018, the company turned 45, and Linda finally decided it was time to shut it down and retire. But it was with mixed feelings. "Over those years we were involved in so many projects. I never could have imagined life being so exciting and wonderful."

CHAPTER 7

OceanWorks: Synthesis of innovation

2001–18

Certainly, the majority of workers in any subsea company spend their shifts in labs, at work benches, running numbers on computers, assembling or testing components, or going to work on or under water. All of that is essential to the production of underwater equipment and the systems that enable them to function. Some are busy devising and testing new products. Others are sniffing out leads that might lead to future contracts or submitting bids.

But at the boardroom level of corporate business, there's often another reality—the financial wrangling associated with company buyouts, takeovers, mergers, subsidiaries, rebrandings or restructurings. Some or all of these can come into play as companies struggle to stay afloat financially, scramble to expand into new markets or even attempt to offload unwanted sectors. Some of these deals leave lasting scars on those caught in the turmoil. For others, it's just another facet of big business.

During the late 1990s, various subsea companies jockeyed to grab a bite of the lucrative oilfield business in various locations around the globe. Companies also schemed to acquire expertise, technology and contacts that might translate into lucrative government or industry contracts. Oceaneering International had already seen huge growth and a number of fractious changes, some of which created hard feelings. Co-founders Lad Handleman and Phil Nuytten both exited the company—Lad in the mid-1970s and Phil in 1984.

Then in 1996, American Oilfield Divers (AOD) of Louisiana engineered a hostile takeover of Phil Nuytten's company Hard Suits Inc. That takeover took a cruel toll on workplace loyalties, resonating to this day in the relationships of those involved: Phil had gifted stock to several employees, but tempted by AOD with offers higher than the market price, enough of them sold their shares for AOD to acquire a controlling stake. Not only did AOD get the shares of stock they needed, but more strategically, they acquired ADS and submarine rescue technology and naval connections. Phil Nuytten, whose company Can-Dive Marine Services had created the submarine rescue system REMORA as well as his invention of the ADS *Newtsuit* technology, still feels like he was thrown under the bus with the hostile takeover. That back-room scheme plumped up American Oilfield Divers' company profile, critically enhanced their position in the Gulf of Mexico, and helped establish the illusion of a high-tech company. Rod Stanley, who had been with Phil's company, took over as the CEO of AOD.

Taking advantage of their new profile, American Oilfield Divers cleverly rebranded itself as Ceanic, a name suggesting the potential of a deepwater off-shore market. Several management shifts happened in quick succession. Rod Stanley left in 1998, replaced by Kevin Peterson. Hard Suits Inc. was now a sub-sidiary of AOD/Ceanic. The new CEO brought John Jacobson into Ceanic to run the Hard Suits subsidiary. That same year, the parent company was purchased by Stolt-Comex-Seaway for a reported $222 million. And that company was soon rebranded as Stolt Offshore.

That acquisition was described as "an over-inflated disaster" for Stolt. Glen Viau, former OceanWorks Inc. COO, adds: "They wanted American Oilfield Divers primarily for its presence in the Gulf of Mexico, so were keen to divest any assets not directly beneficial to those big contracts. Many in the parent company felt that the Hard Suits Inc. subsidiary was competition to Stolt's ships and their saturation-diving work." As a result, the subsidiary was not promoted within the organization. In fact, Hard Suits Inc. was encouraged to undertake a management buy-out, using US Navy contracts and the exper-tise of previous ADS and submarine rescue contracts as the foundation for attracting financing. John Jacobson led the management buy-out that eventu-ally formed OceanWorks in 2001.

Workers in the trenches could only shake their heads as these company trades happened in rapid succession. One long-time employee candidly stated: "As a worker, you just had to laugh every time you got handed a new business card. But it didn't really affect you as long as the paycheque kept coming in."

The quest for a lucrative military contract involved a long process that began companies ago, back in 1992. OceanWorks was eventually successful in securing its first US Naval contract to develop the ADS HS2000. Jim English recalls: "Dan Kerns was a Lieutenant Commander at the US Navy Deep Submergence Unit in San Diego, the military folks tasked with submarine rescue. During their rescue trials in 1992, it had taken nearly 48 hours to remove the fairing so that the hatch on a submarine could be opened. Lieutenant Commander Kerns wondered if an ADS like *Newtsuit*, using an oxy-arc torch, might cut it off quicker."

Jim recalls telling the commander "yes!" and then having to quickly assemble a group to figure out how to do that—and do it safely. Their solution was a tool fitting that could be attached to the *Newtsuit*'s hand pod and hold an underwater cutting torch. In a test tank, with US Naval personnel watching, it cut the fairing off in a mere 15 minutes. As a result, the Navy bought a standard *Newtsuit* for testing, to determine if it could be built to their certification standards.

Contract #1: *HS2000* (HARDSUIT 2000)

In 1996 American Oilfield Divers (AOD) of Louisiana engineered their hostile takeover of Phil Nuytten's Hard Suits Inc. Those employees who stayed with the company after the takeover continued to service existing ADSs and sell new systems to other navies and clients. But the key to bigger money for AOD was certainly the allure of a contract with the US Navy, and 1996 was also the year in which the Navy's *HS2000* program started. That Navy contract work would continue for almost a decade, well into the creation of the Canadian subsidiary of OceanWorks, as described later in this chapter. Despite various company banners, what was most important about the contract was that it provided an eventual route in to the US Navy's much larger submarine rescue program.

After the Navy acquired an off-the-shelf *Newtsuit*, Hard Suits Inc. began consulting with the Naval Coastal Systems Station, as it was called then, to see if such a suit could be certified under the US Navy's stringent SUBSAFE program. They finally agreed that it could, as long as the hulls for the suit were forged rather than cast, thus increasing their strength. In addition, the suit's depth rating was to be increased from 365 m (1,200 ft) to 610 m (2,000 ft). In 2000, aided by Terry Thompson's background as a US Navy SEAL and his knowledge of various company product lines, the US Navy awarded Hard Suits

Inc. a contract to develop a 610-m depth-rated ADS that would meet Navy certification requirements. The *HS2000* became the official name and the project was finally underway

This happened as Stolt decided to divest itself of Hard Suits Inc., so the US Navy design contract began under new independent management but with the same Hard Suits Inc. name. Jim recalls, "The changes going on in our company and the *HS2000* program would soon overlap with the US Navy *PRMS* Submarine Rescue System contract—it was a busy time! We would finish up both projects as OceanWorks International Corporation."

In 2001, two OceanWorks companies were set up: OceanWorks International Inc. (OWI) was headquartered in Houston, and the Canadian subsidiary, OceanWorks International Corporation, or OceanWorks Corp (OWC), was based in British Columbia. Also in 2001, the company renamed the *Newtsuit* the *HARDSUIT*, and OWC undertook upgrades and refits for existing ADS customers. The company then rebranded the suit as the *HARDSUIT QUANTUM*, and several were sold to the navies of Turkey and Russia. OWC also set up an offshore-oil support operation in China. The company continued to expand its submarine rescue capabilities and established an international presence in the marketplace, particularly in Australia and Turkey.

HS2000 systems ready for delivery to the US Navy. Personal collection of Jim English

The company delivered four *HS2000* systems and three Launch and Recovery Systems (LARS) to the US Navy, with the first deep manned dives taking place in 1998, though the *HS2000* wasn't fully certified until 2006. However, the US Navy eventually ended up disbanding their program since the *HS2000* suits did not work as well as expected.

When a submarine has sunk or otherwise become disabled and cannot communicate (a DISSUB), a complex series of notifications goes out, first to the International Submarine Escape and Rescue Liaison Office (ISMERLO) and the international naval community. Next, international submarine rescue systems are alerted and begin a co-ordinated response to get an escape support and rescue system to the location ASAP. Depending on location and depth, sometimes the only hope is a submarine rescue system with a vehicle such as the REMORA or the PRMS, shown here, which can lock on to the submarine rescue hatch and transfer crew members back to the surface, one load at a time. Personal collection of Jim English

Contract #2: Pressurized Rescue Module System (*PRMS*)

After landing the *HS2000* naval contract, Jim Gibson, Terry Thompson and Jim English started on what Jim English describes as "our dragon-slaying hunt" in late 1996 and 1997—with the two biggest dragons they were hoping to bag being NATO and the US Navy. Jim English recalls, "Talking with the US Navy during the *HS2000* program had put us in direct contact with the people that would be heading up their submarine rescue system. Terry Thompson was a key part of this and introduced us to the right naval personnel, including Admiral Joe Krol and Commander Duncan McLean." They wanted to know more about the experience of Can-Dive building the submarine rescue vehicle *REMORA* and how it worked. Jim knew the *REMORA* intimately, having program-managed technical development and execution of the near-impossible deadline of the 1995 project for the Australian Navy that Can-Dive had successfully completed. As a result, Jim spent a lot of time answering the US Navy's questions—information that helped in securing the bid.

Jim Halwachs put together and managed an outstanding team at OceanWorks, eventually delivering a very robust and functional *PRMS*. Upon delivery, Phoenix International and OceanWorks, in conjunction with US Navy personnel, became responsible for operation and maintenance of the system. Personal collection of Jim English.

Jim Halwachs joined Hard Suits Inc. in 1997, bringing his degrees in ocean engineering and 20 years of US Navy service to guide the company through the complexities of a naval bid process and then, hopefully, contract execution. The company worked on amassing information and making plans two full years before the US Navy released the detailed specifications in its Request For Proposal (RFP) of November 1999. Interested parties were given just six months to respond with a detailed formal proposal. Halwachs would be program manager for the project if Hard Suits got the contract.

In September 2000 the company received a two-paragraph notice stating that they had won the contract. It was to build a new submarine rescue system, the *PRMS*, with a vehicle capable of holding 16 people. It would be a second-generation *REMORA*, the original system that Phil Nuytten had invented and patented in 1995 (Patent# 08/561/562 "Articulating Pressure Conduit"). The *PRMS* was essentially Nuytten's *REMORA* system upgraded to go deeper and be larger. Also more complex—and considerably more expensive.

The original bid was approximately US$24 million with delivery of the submarine rescue system in four years. This budget and timeline would grow to over $75 million and nearly seven years. The increase wasn't surprising given the rigorous standards and the number of meetings, design reviews, quality audits and complex certification issues.

Finally, the Canadian subsidiary OWC completed a robust and functional vehicle. Shallow-water trials began at Indian Arm and then moved to Jervis Inlet for deepwater testing. The PRMS was shipped to San Diego in 2007, and the vehicle was officially christened FALCON in 2008, replacing the US Navy's DSRV MYSTIC. It was a huge coup for a small company to win and execute a multi-million-dollar, multi-year contract over Lockheed Martin, a major US defence contractor and OceanWorks main competitor on the bid.

Mavis Mitchell recalls the mountain of paperwork associated with the PRMS contract. "The paperwork just buried us. We built and delivered the PRMS, but it took two more years to close the paperwork," she laughs. "The US Navy is a tough customer but boy, you sure learn a lot working for them. They're exacting, and so very, very careful with everything that they do. I have a lot of respect for getting the chance to work with them and learn from them."

The history of the two OceanWorks

These two long-term naval projects began under one company banner and continued under new ones, with the names of key players in flux as well. In 2001 OceanWorks International Inc. (OWI), headquartered in Houston, was spawned from the Hard Suits/AOD/Stolt evolution. Many considered the US ownership and structure of the company lopsided, given that the majority of the business revenue and technical talent was centred in Canada and brought in through the Canadian subsidiary OWC. But no Canadian investors surfaced. Jim English became VP, general manager and technical director of the Canadian entity; Jim Adamson was VP Remote Systems, based in Florida. The outside investor was the Wedge Group of Houston.

The intention was that the Houston facility was to support itself with oilfield work in the Gulf of Mexico; the Canadian subsidiary would get its main revenue from ADS sales and submarine rescue projects, having acquired Hard Suits Inc. and its submarine rescue technology and existing contracts from Stolt. Somehow, that financial balance never quite worked out. "Houston never made money," Glen Viau states of the parent company. "In fact, they bled us dry year after year."

The formation of OWC was tumultuous. The small Canadian company struggled along with only \$10–\$14 million in annual revenue, compared with industry giants like General Dynamics (\$36 billion annual revenue) and Lockheed-Martin (\$54 billion). When OWC won the contract for the US Navy's PRMS program in 2001, the Canadian subsidiary could only claim around 25 staff. At the height of the submarine rescue boom, the company had grown to over 145 employees, along with independent contractors such as ISE, which fabricated the control systems.

From 2001 to 2011, OWC matured. The Navy contracts not only provided big money, but forced the company to learn how to manage and execute projects. However, the key to the company's success was its talented and motivated people, who just ate up technical underwater challenges. Although OWC was always under-funded and struggling with cash flow, the company continued to put out excellent products and services. The company was also small enough to be responsive to short schedules and one-off crazy projects, and had the talented people to pull it off.

In 2005 Rod Stanley had joined the company as CEO of OceanWorks Inc. in Houston. The aim was to develop a strategic plan for the company. He had already done this once before with a large ROV company, but the hoped-for redevelopment strategy never materialized for OceanWorks. That same year Jim English stepped aside as general manager to become regional VP. Glen Viau became general manager and later COO. Engineering manager David Lo deserves credit for developing an engineering team at OceanWorks that included a diverse mix of mechanical, electrical and control system engineers, all of whom greatly enhanced company capabilities.

In 2005 and 2006, OWC diversified and got into the undersea observatory business. Along with Global Marine, OWC was awarded a contract by the University of Victoria for their VENUS observatory system. Ocean observatories (described in Chapter 13) involve a network of hybrid power and fibre-optic cables running along the seabed and connecting subsea nodes. Each node bristles with various sensors that collect continual data on a variety of aspects of the ocean environment. These underwater observatories permit researchers

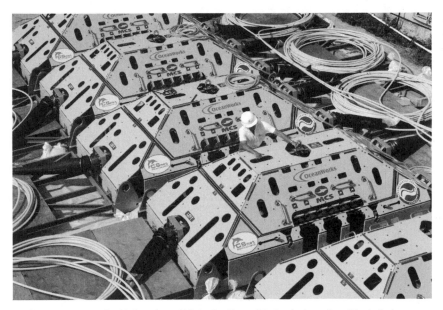

Under contract to the University of Victoria, OceanWorks designed and installed *VENUS*, the world's first cabled observatory, beginning in 2006. Subsea fibreoptic networks and observatories span thousands of kilometres on the seafloor to depths of 6,000 metres, collecting data on physical, chemical, biological and geological aspects of the ocean over long time periods. Personal collection of Jim English

to receive data anywhere in the world, in near real time, 24/7/365. Generally, data from these observatories and networks is openly available and free.

VENUS Phase-I went live on February 8, 2006, and Phase-II in August 2007. In that year the University of Victoria established Ocean Networks Canada (ONC) to manage the *VENUS* and *NEPTUNE* observatories. Although OceanWorks did not win the bid for *NEPTUNE*, the largest cabled ocean observatory, the company did receive a subsequent contract to provide junction boxes for the scientific instrument interfaces and subsequent additional systems, upgrades and technical support. On December 8, 2009, *NEPTUNE* came online with an 800-km-long (500-mi) seafloor cable running from a shore station at Port Alberni across the continental shelf into the deep sea. Regional underwater observatory equipment and services became a major product line for OWC, and the company built cabled observatories for a number of countries and scientific institutions.

In 2008 the company set out to find bigger facilities with greater office space and fabrication space, finding the ideal place in Burnaby. Once OWC located a suitable facility and modifications were made, they brought in their hyperbaric test chamber and shallow-water test tank, then opened the doors. The new facility would make a positive impression on big clients like British Petroleum.

In spite of ongoing management upheavals, OWC really spread its wings in the international marketplace until about 2014. The huge US Navy contracts not only brought almost CDN $100 million into the company, but allowed OWC to expand engineering capability as well as project management and quality control. The Canadian subsidiary took on other contracts as well—a marine well containment project, a research and development project for British Petroleum, a big subsea rescue contract in Turkey, and more cabled underwater observatories.

"In 2011 we had a record annual revenue and over a hundred people on the payroll," Glen Viau recalls, "but we had no money! Year-after-year losses in Houston were extremely detrimental to OWC's cash position in Canada. It came to a head in Christmas 2009/New Year 2010, when OWC managers had to make personal loans to cover payroll. In early 2011, the company again had to take emergency loans from shareholders to cover payroll." Frustrated, Glen left the company in November 2011, going to work in Houston's oil and gas sector for the next four years, returning to Canada when the oil and gas market collapsed worldwide. Jim English had already stepped back from senior management in 2005 and finally resigned in 2013.

Many felt it was obvious that the Canadian arm of OceanWorks was basically continuing to fund the Houston company. It was also becoming apparent that attention to customer satisfaction was eroding and company morale was slipping. Furthermore, there was a lack of communication, and the leadership and vision from Rod Stanley never did materialize. Communications between North Vancouver and Houston were often antagonistic. The personnel changes, contract performance issues, continuing lack of communication, problems with cash flow and investment partners all culminated in several break-even and unprofitable years.

A bad end

This is where the story gets even more convoluted, as big international business and new majority stockholders took over. Various company ownership changes resulted in confused business plans, and management direction became truly snarled. Eventually, the Canadian and US federal governments stepped in. Opinions and accounts vary, so it's difficult to figure out exactly what happened.

Briefly, in 2014, a US company by the name of Revere Merchant Capital bought OceanWorks International Inc. and its Canadian subsidiary OWC, becoming the new majority shareholders. They quickly reached a decision to sell the Canadian arm of the business, along with its ADS, ocean observatory and submarine rescue technology. It's possible their rationale was to retire company debt so they could continue to operate the Houston office. It's also possible that they just wanted to get their money out of a flagging investment and the Canadian subsidiary was the only thing of value. Whatever the reasoning, it was the ultimate disastrous management decision for Canada's OWC.

CEO Rod Stanley left the company in 2016, and Revere Merchant Capital negotiated the sale of OceanWorks Corp, truly the heart and soul of the company, to Highlander Ocean Engineering Technology Development Co. Ltd. of China. Highlander was keen to expand into the marine technology business, so the previous year they had purchased Laurel Technologies. That company had been the Chinese agent for OWC for many years, promoting submarine rescue and ADS to the Chinese military. As a result, Laurel Technologies knew and recommended Glen Viau for the position of managing OWC. So, when the initial sale of OWC to Highlander went through, the company hired Glen Viau as president in 2016.

Serious problems with the sale soon materialized for the new Chinese owners, due to a dispute over the cost of OWC and the fact that the acquisition deal was completed before approval under the Investment Canada Act. In December 2016 the new Chinese owners were informed that the Canadian government was going to conduct a National Security Review. That review process lasted about five months, during which time OWC was mainly focused on a major submarine rescue opportunity with the Australian Navy.

OceanWorks teamed up with the US company Phoenix International for the Australian bid, and by early 2017 the team believed they were the frontrunner. Then in May 2017, the Canadian government issued a divestment order requiring Highlander to sell its stake in OceanWorks within 180 days. The timing of this divestment order would prove disastrous. In August 2017 the OceanWorks/Phoenix team were notified that they had been selected as the preferred bidder for the Australian Navy contract and negotiations began. Unfortunately, the divestment order eventually forced OceanWorks to sell its submarine rescue technology to Phoenix. As a result, Canada's leadership in submarine rescue technology ended. OWC's cabled observatory business was sold to Ocean Networks Canada, and a new company formed by Glen Viau acquired the OceanWorks name and the atmospheric diving system business.

The Canadian National Security Review and the subsequent divestment process also resulted in significant legal jeopardy for Glen Viau, who was indicted by the US Justice Department for making a false statement in the voluntary disclosure. The case was settled in December 2019 with both Glen and the former company agreeing to pay a fine. Despite the nasty legal snarl at the end, those workers who were part of the OceanWorks team in Canada certainly remember the company fondly. Mavis Mitchell and her husband, designer Curt Schmidt, were hired as a team, starting with Can-Dive Marine and Hard Suits Inc., then staying on once control of the company segued to OceanWorks. "Even though the company was dissolved in 2018, we still get together regularly," she says. "OceanWorks was a great place to work. Everyone was really good to one another. And everything was a group effort. All of us were really sad when we lost our work there."

The complicated end of this Canadian subsea company brings up a lot of questions, with widely varying answers. What is known is that Canadian submarine rescue technology was acquired by the United States at bargain-basement prices. What is also apparent is that prior to this, Canada had become a world leader in submarine rescue, first with Can-Dive and then with OceanWorks. After the dust settled on the sale of assets and the court proceedings, Canada was left with nothing.

CHAPTER 8

International Submarine Engineering (ISE) and James McFarlane

1974–present day

James McFarlane Sr. (Jim) was not the only skilled employee to leave International Hydrodynamics in the early 1970s. Many felt that HYCO's financial structure was already shaky and disagreed with the new management's direction for the company, or were frustrated by HYCO's convoluted decision-making procedures. But for Jim, a particularly dramatic event kickstarted his thinking. It came at 3 a.m. when the RCMP arrived at his door with a Subdown/Subsmash Notification—the emergency alert at the time for any submarine or submersible in serious trouble. In this case, *Pisces III* had gone down off the coast of Ireland on August 29, 1973, with Roger Chapman and Roger Mallinson aboard. They were suddenly stranded with limited oxygen at a depth of 1,575 ft (480 m), which is still a record for a manned submersible recovery.

That emergency notification immediately mobilized a broad array of HYCO vehicles and talent, including Al Trice, who was travelling in England with his wife and young daughter, and also Mike Macdonald, Jim McBeth, Bob Starr, Bob Holland, Al Witcombe and Steve Johnson, who were burying cable with *Pisces V* in Halifax. The US Navy readied their new ROV, called CURV, which had already successfully recovered a nuclear bomb lost off the coast of Spain. Working behind the scenes, Jim facilitated military transport of crew, vehicles and supplies from Canada, while Al Trice became second-in-command to Britain's Peter Messervy.

The harrowing rescue, including the relentless nasty weather, is relived in Roger Chapman's book *No Time on Our Side*. After its publication, many felt the Canadian contribution was not accurately credited, so Jim collected first-hand accounts from those involved and wrote up "The Canadian Contribution to the *No Time on Our Side* Story." It details the crucial role played by the Canadian Air Force, the Canadian Navy, Teleglobe Canada, and the HYCO crew led by Al Trice and Mike Macdonald in this gruelling and ultimately successful operation.

Indeed, it was Mike Macdonald and Jim McBeth aboard *Pisces v* who eventually found and attached a locator line to *Pisces iii*, and stayed with the downed submersible on the ocean bottom. Then *Pisces ii* and the US ROV *CURV* (repaired and re-deployed) were finally able to attach lift lines that held and got *Pisces iii* back to the surface just minutes before the crew's oxygen supply ran out.

Once normal work resumed, Jim began to actively consider the possibilities of remotely controlled, tethered, unmanned work vehicles, particularly now the rest of the world was aware that an ROV had been a key part of the successful recovery. "That's the way of the future," he stated bluntly. "That's the direction this industry is going to go." He was right. And much of the rest of the underwater world would soon reach the same conclusion. The significant cost of support ships for manned diving and also the introduction of high-density electronics signalled the end of submersibles for oil work and cable laying. Unfortunately, the management at HYCO disagreed.

As a result, Jim left the company, taking with him his engineering expertise, his technical knowledge, and his connections to and understanding of the military world. On August 13, 1974, he and a funding partner founded McElhanney Offshore Surveying and Engineering Ltd. In 1977 he changed the name to International Submarine Engineering Ltd., known worldwide as ISE.

Jim is synonymous with ISE. Both the man and the company could boast of stellar achievements if they were that way inclined. But they're not. Born in Winnipeg, raised in Clandeboye, Manitoba, and Chilliwack, British Columbia, Jim attended the University of New Brunswick, earning his degree in mechanical engineering and going to sea during the summers. He is proud to have started in the Canadian Navy as an ordinary seaman, a naval airman, "at the level of washing windows," he laughs. The Navy also invested in his education, later sending him to MIT for two and a half years, where he earned two Master of Science degrees—one in naval engineering and a second in naval architecture and marine engineering. He saw active submarine duty in England and was also in charge of constructing the Oberon Class of diesel-electric submarines—the *Okanagan*, *Ojibwa* and *Onondaga*.

Some 18 years later, Jim retired from the Canadian Navy with the rank of lieutenant commander. After overseeing the completion of the SDL-1 project that International Hydrodynamics Co. Ltd. (HYCO) fabricated for the Canadian Navy, he joined that company as vice-president of Engineering and Operations. In 1974 he left HYCO and founded the company that became International Submarine Engineering. Now in his mid-80s, Jim can still salt his vocabulary with words like "chemosynthetic lithotrophic thermophilic heterophiles." At the end of one of our talks, he handed me an ISE baseball cap emblazoned with a typical McFarlane motto that speaks to his attitude about the accomplishments of his illustrious career—Git 'er Done!

The wisdom of building ROVs

Jim McFarlane and his company would soon become part of the incredible undersea revolution of the mid-1960s. Prior to that time, divers had only ventured underwater to about 300 ft (90 m) and, in early submersibles such as the Bathysphere and the Benthoscope, down to about 4,500 ft (1,370 m). Then in 1960, Jacques Piccard and US Navy lieutenant Don Walsh descended more than six *miles* to the bottom of the Mariana Trench in the Pacific, the deepest place in the world (35,798 ft / 10,911 m). That event triggered the development of subsea exploration, including a variety of manned submersibles, some of which Jim worked on at HYCO. Now he was determined to build the next wave of subsea craft. Those would be vehicles equipped with cameras and manipulator arms that could do a diver's job at greater depth but with far less risk to human life and less expense. From the get-go, he set out to be different.

Those early years were anything but easy. A September 1986 *Canadian Business* article noted that Jim soon discovered that other companies had also decided to build unmanned ROVs. And many of those other companies had access to large capital bases. ISE did not. As he told the magazine, "At that point you either stop or you fight. So it was a fight." And he adds, "We built boats, fish-gut tanks, trays for Safeway [supermarkets]. Anything to make a nickel to

ISE's *TROV* was the first commercial ROV to operate in the North Sea, monitoring rock dumping on the Piper Alpha line. Jim McFarlane says that these new vehicles "kicked the pants off manned submersibles." *TROV* would go through several changes and upgrades, including *TROV N* which was a 40 hp 1,000-m/3,280 ft torpedo-recovery vehicle built for the US Navy. Collection of ISE, Ltd.

survive." With characteristic blunt understatement, he sums up those initial years, saying, "It was reasonably tough." However, his single-minded vision and determination would eventually push ISE to world-class prominence.

The distinct financial advantages of ROVs can be summed up succinctly— "No payroll, no benefits, no cook, no crew." And "less insurance hassle" could certainly be added to that list, since insurance companies have always been wary of manned things operating underwater. That was another reason why the offshore oil and gas people started switching over to ROVs—there wasn't any human life at stake.

Jim committed his fledgling company to building ROVs. In 1974, after working on a prototype in Mack Thomson's well-equipped basement workshop, International Submarine Engineering Ltd. produced its first ROV, called *TROV* (for Tethered Remotely Operated Vehicle). It was built for Canadian inland waters and in 1975 discovered and explored the warship wrecks in the Great Lakes from the War of 1812. A few years later, ISE marketed its visual ROV *TREC* (for Tethered REmote Camera).

Mike Macdonald, who became ISE's first executive vice-president, recalls those early days: "I decided I was going to leave HYCO, too, and in 1977 James Sr. asked me to join him. He interviewed me in a room he was sharing with a very large Xerox machine belonging to McElhanney Offshore Surveying and Engineering. One of their principals put some money into Jim's new company so we were able to use their office downtown. Jim was busy writing proposals that were all about ROVs. That was the way to go."

TREC soon replaced divers who had previously done walking inspections of oil and gas pipelines in the Gulf of Mexico. Collection of ISE, Ltd.

Mike also remembers taking *TROV 1* over to the North Sea in November 1977 to inspect the Piper Alpha pipeline: "It was the first time an underwater inspection had been done without a diver walking in front of the support boat." In fact, *TROV* was the first commercial ROV to work in the North Sea, ushering in a new era of subsea work capabilities. Not to be left behind, *TREC* was the first ROV to do a pipeline inspection in the Gulf of Mexico. Soon those early ROVs replaced the commercial divers that had been servicing well sites. And insurance companies were more than happy to replace people working underwater with less-vulnerable machinery.

ISE went on to produce several *TROV* versions, including *TROV N*, which was an upgraded torpedo-recovery ROV with 40 hp and a 1,000-m (3,280-ft) diving depth, as well as manipulators, torpedo clamps and dredging equipment. In essence, it took over the job that *Pisces* had originally performed at the torpedo range in Nanoose Bay. That switch was typical of how quickly

ROV systems replaced submersibles as well as commercial divers. *TROV N* was the early predecessor of ISE's family of *HYSUB* ROVs. Other *TROV* versions for oilfield work soon followed.

In 1979 Mike Macdonald, alongside pilots working for MARTECH International at the controls of the three *TREC* ROVs that MARTECH owned, was able to assist the famous Red Adair in capping Mexico's *Ixtoc 1* well blow-out, which was gushing 30,000 barrels of crude oil per day into the Gulf of Mexico. During one dive, the ROV was sucked up and through the vortex of the burning wellhead fire. Macdonald recalls, "We got right over the blow-out itself and the ROV got blown out of the water. We actually got the vehicle back but it was pretty badly damaged. Twenty-four hours later we had it in the water again."

In that year Doug Huntington joined ISE and worked a couple of years as sales manager. He recalls that other companies also had the vision of ROVs taking man out of the ocean, "but Jim McFarlane was right on the edge. He was one of the top companies, along with Perry Oceanographics in Florida, Slingsby Engineering in the UK, and Ametek Straza in California. Hydro Products also built a successful observation vehicle, the *RCV 225*, but ISE built a competitive vehicle, the *DART*, for about a tenth of the price. Other companies came on later as the market grew, but in the beginning there were only a handful of early inventors of ROVs. Essentially, they built them from scratch."

He adds, "Jim McFarlane was quite a workaholic and burnt me out at the time. But he was amazing! He would often send off a proposal for an unsolicited piece of equipment to a company when they didn't even know they needed it. He did that with the Canadian Hydrographic Service, and that craft ended up being ISE's semi-submersible *DOLPHIN*."

As production began to ramp up, ISE worked closely with other companies to access needed expertise and equipment. Among these were RMS Industrial Controls Ltd., a BC company that provided telemetry and control systems, and Robotic Systems International (RSI) in Sidney, BC, which manufactured robotic arms, hundreds of which sold worldwide. International Submarine Technology Ltd. (IST), based in Redmond, WA, was a subsidiary that produced advanced seabed imaging equipment.

ISE soon realized that opening a Houston office was the only way the BC-based company was going to get any business in the Gulf of Mexico, so the subsidiary company ISE Gulf was set up in the late 1970s with Mike Macdonald as president, building on his years working there for HYCO earlier in the decade. Now be brought on board people like Doug Hernandez, who worked for the company for many years and then became "Mister ROV" for British Petroleum.

At ISE Gulf, Mike became aware that oil companies did not want to buy equipment such as ROVs themselves; instead they preferred making short-term contracts with service companies that already owned the equipment. So he invested a lot of his time in Houston building relationships with diving/service companies. Essentially he was a sales engineer, though he hated cold-call marketing. But he realized that marketing was often about an engineering problem, a service problem or an operational problem. And he liked *those* calls. "I quickly found out that if you went out troubleshooting more than once, the company was glad to see you and interested in more work with you." As a result, International Submarine Engineering started getting orders.

"ISE's new *HYDRA* ROVs really made our reputation with Oceaneering. In the early 1980s, *HYDRA 2500* set a world depth record for oil drilling support, diving 7,000 ft off the New Jersey coast." Admittedly, ROVs were not as reliable in those early days, so Oceaneering's solution was to buy one handling system, one console and two vehicles, just in case. "Of course, soon companies wanted to go deeper and have more horsepower, so the ROVs started getting bigger and beefier."

TRAPR (for tethered remote autonomous pipeline repair) certainly was bigger. In fact, its size was comparable to a 5-ton box truck. The mission for this large specialized ROV unit was conducting pipeline repairs remotely on the seabed. It actually had another smaller ROV that flew out of it to monitor and assess the pipe.

ISE developed an interesting sideline—not only did they manufacture ROVs, but they also trained divers from other companies to operate them. Mike explains: "Divers make very good ROV pilots because they have good situational awareness and often work in poor visibility. You have to be able to look at something you can't hardly even see and think, 'It's over there, so if I turn this way, I'll be right at it.'"

When foreign countries purchase a vehicle from ISE, they buy the whole kit—the vehicle, the console, the winch, and the proprietary software that

Don Muth credits Jim McFarlane, Mike Macdonald and Al Trice as significant mentors: "I continued to learn from their experience. This was and is ISE's company culture, so we pass that learning on while continuing to learn ourselves."
Personal collection of Don Muth

the vehicle runs on. Once the vehicle is ready for the test tank, generally the country buying it sends over a crew to be trained on-site in various classes. Some are pilots, others are technicians; many are both.

"The training usually starts in the final stages of the vehicle build so that the new crew can become intimately familiar with the system," Don Muth explains. He's an instructor as well as a senior technical adviser with 38 years of experience at ISE. "There's also classroom work from experts in various areas: electrical, software, and hydraulics. Then there's hands-on work in some of the final assembly, followed by the factory acceptance test. Then it's on to the pool and sea trials. Many times, ISE has also had two or three of our people go on board ship for the first customer dives. They do additional training and provide field support so that the customer's first jobs go smoothly and successfully."

Most crews in training come to ISE with a working command of English; if not, they bring a translator. And most have some experience because their company has other underwater vehicles. "That makes a big difference. But even when they know how to operate a 60-horsepower vehicle, there's still a learning curve to move up to one with 150 horsepower. So there's always a need for training."

Don Muth learned the value of training and mentors himself, beginning in March 1982. He got laid off from a job during a recession, but fortunately had a degree as a mechanical technologist from BCIT. Within ten days he was working at ISE and has never looked back: "The excitement of working at the cutting edge of technology and seeing a job through to completion makes it a pleasure to come to work."

Expanding the company's mandate

The oil and gas industry isn't just about drilling and completing wells—just ask any support company. There's also an enormous range of services involved that includes laying, servicing and inspecting pipelines. Trans-Atlantic telephone cables also needed similar services. And cable-laying contracts increased as communication demand began to grow, first using co-axial cables with wide bandwidth and repeaters, then later fibre optics. But there can—and will be—problems with any cable system. This unhappy situation is known as "downtime" in the industry. Generally, it's because some commercial fishing vessel's net has accidentally snagged the cable and damaged or cut it. Regardless of how it happens, downtime is always expensive and inconvenient. As a result, in the shallower waters of the continental shelf it became worth the cost for communication companies to have ROVs bury their cable in the seabed, rather than just lay it on the bottom. And that created another business opportunity that the company was quick to address.

As ROVs took over more and more of the support work for oil and gas companies during the 1970s, ISE became known for its successes in the oil patch. That reputation led to some interesting alternative contracts, many of which involved ISE manipulators. In 1979, the Three Mile Island Nuclear Generation Station in Pennsylvania suffered a partial meltdown of reactor #2 and a subsequent radiation leak. It proved to be the most significant accident in the history of US commercial power plants. ISE was asked to do robotic work on the clean-up of the highly toxic environment, using a manipulator to go right inside the reactor core to assess the damage. To do so, the company developed a manipulator whose hydraulic system used borated water instead of the normal oil. Don Muth explains that using water-plus-boron actually absorbed some of the radiation, so it was safe to operate the manipulator system in that highly radioactive environment. It's the same borated water used when storing nuclear fuel rods in a core cooling system. "It was fascinating for us technical guys."

While most Canadians know of the Canadarm, an integral part of the International Space Station, few would also know that ISE built several related manipulators for the Canadian Space Agency. These on-the-ground mechanical arms served as training devices so that astronauts could practise various

ISE has produced over 400 robotic manipulators. Manipulators are often the unsung but essential part of any manned or unmanned underwater vehicle, allowing the pilot to perform a variety of tasks. Work-class ROVs are usually equipped with dual manipulators. There's often a 4-function (4F) arm and a more dexterous 7-function one. The "hand" or end-effector on a manipulator can be interchangeable. Collection of ISE, Ltd.

tasks before going up in space and using the actual Canadarm (which, because it was designed to be used in zero gravity, was too heavy to work on Earth). ISE had seven or eight people working on this "sideline" project for three years. In 1985, an AT&T *SCARAB* submersible, equipped with ISE-made robotic arms, grabbed the black-box flight recorders from the sabotaged Air India flight that crashed into the Atlantic.

However, ISE also won some contracts that had nothing to do with manipulators, ROVs or subsea activities. Shell Oil Products spent a great deal of time and energy exploring the idea of auto-refuelling cars, and for over three years ISE worked on the development and prototype build of an automated auto-refuelling robot for the company. However, the reality was that no matter how bad the weather, people were still more comfortable refueling their own vehicles, so the project was eventually dropped.

ISE also worked on projects for the mining industry, another deep, dark and dangerous environment where it is safer to get humans out of the loop.

Two notable projects included the automation of a potash reboring mill that changed a purely manual material conveyor operation to a fully automated one, removing personnel from a high-risk environment. A second project suited ISE's ROV technology in a complementary way. It involved the design, development and testing of a narrow-vein mining machine, dubbed "the mini mole," to access thin veins of high-grade oar that were not economically viable to mine in a conventional way.

<hr>

The old adage "never put all your eggs in one basket" is good advice for any business, above or below the water. Certainly, when times in the oil and gas industry were good, the money was very, very good. But success in the oil patch has always been tied to the price of oil. And the same is true for ROV sales. In the early 1980s there was a downturn in oil pricing that proved to be crippling for many companies. "Jim saw this coming," Mike Macdonald states. As a result, ISE also began to focus on government work, beginning in the late 1970s. On February 9, 1982, the company launched International Submarine Technology Canada Ltd., with a legal name change to ISE Research Ltd. (ISER) in 1984. The specific aim of this new entity was securing government and military contracts with unique technological systems.

ISER's first government contract (1981–83) was to construct a remotely controlled, semi-submersible hydrographic survey vessel. DOLPHIN is the short version of Deep Ocean Logging Platform Instrument for Navigation (or Deep Ocean Logging Profiler Hydrographic Instrumentation and Navigation). Able to travel at 15.5 knots (29 kph), the vessel is radio-controlled and draws air for its diesel engine through a periscope.

Without intending to, Al Trice became involved in the DOLPHIN project. In early 1983 he was just completing a short-term contract with ISE to write a proposal for the upgrade of the SDL-1—the vehicle that HYCO had built for the Canadian Navy with Jim McFarlane as military supervisor. (The proposal was successful and eventually ISE would do four refits of the SDL-1.) When Al was finishing up the proposal, he got an invitation from James Fergusson, then VP for ISER, to accompany him to Indian Arm, where sea trials were underway for the early DOLPHIN. "I wasn't particularly interested, but once I saw it in

Beginning in 1981, ISE developed its semi-submersible *DOLPHIN* for military applications. Collection of ISE, Ltd.

operation, I was totally fascinated," Al recalls. Fergusson offered Al the opportunity to run that program and he did so for several years. Six *DOLPHINS* mapped Canada's undersea continental shelf. And during the 2010 Olympics, the Canadian military used the semi-submersibles to search Vancouver harbour for mines.

Despite the early-80s downturn in oil pricing that stifled sales of remotely operated vehicles, ISE's ROVS were busy garnering "firsts." In 1983, the *HYDRA 2500* set a 7,000-ft (2,134-m) depth record for oil-drilling support. A year later, with Craig Bagley as chief ROV pilot, along with Oceaneering crew, the company used its *HYDRA 2500* to set another oil patch record, for the first deep AX-ring change-out, at 7,002 ft (2,135 m). That technical jargon refers to the metal gasket that seals the wellhead at the BOP stack to the main riser pipe at the sea floor. For a change-out, the ROV must first take the AX ring and removal tool to the sea floor, then unscrew the joint and position the removal tool (puller) in the AX ring, and finally winch up the removal tool (a job requiring 15 tons of force). Next, the ROV must carefully position the new AX ring in the BOP stack and guide the main riser pipe into alignment, carefully avoiding any damage to the new seal ring. This last step is a critical one since the joint must contain pressures of more than 20,000 psi (1,400 kg per sq cm).

Switching vehicle types and moving from the oil patch to the shopping mall, in 1985 ISE got a most unusual contract to build four submarines that would only submerge 15 ft (4.5 m). They were part of the Phase Three expansion of the West Edmonton Mall. ISE fabricated and delivered four 38-tonne HO (for Human Occupancy) vehicles. Each sub accommodated 22 passengers and cruised 15 ft down in an enormous man-made lake. It was often noted that, with four submarines, the West Edmonton Mall had one more than the Canadian Navy. After 20 years of service, the rides were halted, but the subs weren't scrapped until 2012.

International Submarine Engineering's initial goal was to build ROVs and they continued to do so. Most of them were intended to meet the emerging demands of the offshore oil and gas sector. But Jim McFarlane was certainly open to other options. Expo '86 provided an opportunity to showcase BC technology on several fronts. One event involved a satellite hook-up between China and the World of Construction exposition at BC Place. For the televised event, a "slave" robotic arm was successfully activated by remote control from Beijing, enabling a high-tech "handshake" in Vancouver. This space-age razzle dazzle was intended to let China know of Canada's considerable construction capability. The robotic arm, developed by ISE and RMS Industrial Controls of Port Coquitlam, was identical to that used to gather debris from the ocean floor after the *Challenger* space shuttle disaster earlier that year.

Jim McFarlane also vividly remembers a chance encounter during Expo '86 that led to the development of the scientific exploration ROV *Ventana*, built for David Packard in 1987. It began when Grace McCarthy, a Member of the BC Legislature tasked with the Economic Development portfolio, contacted Jim. "She wanted a display of BC trades, so requested that ISE bring one of our vehicles. I took the *Researcher*, our trials vessel and testing platform built the previous year, and one of our near-surface vehicles and had a little stand at Expo. Not many people stopped, although the King of Sweden came by. Then one day a guy by the name of Derek Baylis came along. He asked me what I did and I told him I built underwater vehicles. At first he was skeptical, but as we talked he figured out that I knew a lot of stuff. He listened and then said, 'Oh, Dave would be interested in this.' So I invited the guy out to our shop and showed him around. Again, it was, 'Oh, Dave would be interested in this.' About two days later, the same guy calls me from California and says, 'Dave and I would like to come up.' I didn't know who in hell Dave was, but I offered to pick them up at the airport. 'Not necessary, we've got our own plane and we'll be there tomorrow.'

"The next day this long car drives into the ISE yard and inside was David Packard of Hewlett Packard, and also the founder and funder of the Monterey Bay Aquarium and Research Institute. We took him out on the *Researcher* and demonstrated an ROV. He ordered one on the spot!" That commission became the ROV *Ventana*. ISE modified the ROV design to fit with Packard's vision of underwater exploration. But strangely, Packard was definite about starting with a shorter tether. Jim explained that a much longer one was available, but Packard cut that discussion off, saying, "When I've seen everything at this depth, I'll buy another umbilical. Get it?"

"Yessir. He knew what he wanted, and he knew what he was talking about."

On one trip to ISE, David Packard also met Jim's son, James A.R. McFarlane, who was back from overseas after working for several years in the oil and gas industry and then doing military mine countermeasures with the US Navy. Again, Dave Packard made an immediate decision: "Can you come work for me and pilot this ROV?" James A.R. immediately accepted the challenge. Like his father, James A.R. thought Packard's decision to let his scientists only use a 2,000-ft (610-m) tether, when the ROV was rated to 6,000 ft (1,830 m), seemed very puzzling. "But it became very apparent that *he* was the smartest guy by limiting us depth-wise. Instead of following the conventional thinking of 'deeper is better', we could only research in the upper water column. As a result, we made a new discovery *every day* during those first years."

Jim Sr. made regular trips to Monterey, advising about underwater science and talking about the physics of subsea work, with Packard always in attendance. "That man exuded capability. He had a vision plus determination plus the bankroll. Over the years, his ROV studied Monterey's deepwater canyons intensely and discovered all different types of animals that nobody knew

Custom designing a vehicle such as *Ventana* to carry out its intended mission is a critical part of what International Submarine Engineering does. "ISE doesn't build a hundred identical vehicles," James A.R. notes. And his father would add, "Everything we do underwater is mission-driven." So the attributes of the vehicles that come out of those different experiences are generally unique as well. Collection of ISE Ltd.

were there before, including chemosynthetic lithotrophic thermophilic heterophiles! The entire experience was an awakening—for all of us. It opened up a new world of scientific work."

The 2-ton, 10-ft-long (3 m) *Ventana* received a complete refit in 2017. Over the years, it created a remarkably safe and successful dive record for the Monterey Bay Aquarium Research Institute (MBARI). It's worth noting that the *Ventana* program was also the first fully committed ROV science program, developing the tooling and operational methodology that other programs copied.

During the 1980s and '90s, ISE produced several vehicles specifically designed for scientific work. *ROPOS* (the Remotely Operated Platform for Ocean Science) was one of them, built in 1986–87. As the long form of its name suggests, this subsea robot is equipped with sampling tools, hydraulic thrusters and wide-angle cameras, as well as two powerful and agile 7-function manipulators with interchangeable tools (see also Chapter 13).

ROPOS was large—essentially the size of a compact car—and far from flashy, but it proved to be extremely reliable. It was capable of deep diving for days at a time and was able to conduct deep-sea hydrothermal vent exploration or deploy and maintain seabed observatories. Its three general tether configurations (1,000 m/3,280 ft, 3,000 m/9,840 ft, and 5,000-m/16,405 ft) allowed a variety of operational goals and requirements. Over the decades *ROPOS* and its successor have been at work, both the ROV and the company have benefited from collaboration with leading ocean scientists, engineers and students, particularly when the re-configured *ROPOS II* was built in 1997.

Military contracts also continued during the 1980s and '90s, and up to the present (many are classified so can't be discussed). Design work on the *Aurora*, an actively stabilized towfish, began in 1997. The towfish was designed to fit under variants of the semi-submersible *DOLPHIN*, and later the upgraded *DORADO*, carrying mine countermeasure sonars. The upgraded *DORADO*, bigger and heavier than its predecessor, features counter-rotating props to remove torque. The mast stands up around a pivot point, and when underway only a couple of the fairings on the mast stick out of the water. The fairings on the front stainless-steel stay wire reduce drag, which can be significant enough that the mast would fold back on itself. *DORADO* is powered by a 425 hp 6-cylinder twin turbo diesel in the aft hatch section, with all the machine's telemetry in the forward section. When submerged, the dive-control planes on the front pins provide a very stable platform for doing high-speed, double-stabilized sonar surveys. The value in this, as Don Muth explains, is that it decreases the time required to do the survey as well as decreasing the time needed to post-process the data, since there are fewer pitch, yaw, heave and velocity changes to correct for. The end result is more timely data for decision-making, an especially important feature for military needs.

A subsequent update to the *Aurora* towfish integrated new sensors and navigation equipment. Vince den Hertog, now with Robert Allan Ltd., handled the upgrade. This new generation of commercialized *Aurora* towfish can operate at speeds up to 12 knots (22 kph) and is described as having an "actively stabilized, modular design." The term "active" means it can maintain horizontal position (controlling the pitch, yaw and heave axes) and depth, as well as avoid obstacles. Don Muth explains that active control for sonar is a very attractive feature since sonar only moves at the speed of sound, so an unstable platform blurs the data—like the blur in a handheld photograph taken in low light with a long shutter speed. But just as a tripod-mounted camera results in a sharper photo, so a stabilized sonar platform allows a clearer picture of what's on the bottom. Each towfish possesses its own controller, and the hull can accommodate wet or dry payloads. Its active tail planes help control attitude and stability, and a mounting array on the bottom can accommodate a wide range of sonars.

The growing threat of new-generation mines that are more difficult to detect triggered the need for new mine-hunting systems with improved safety and efficiency. The upgrades to both ISE's DORADO and *Aurora* towfish helped to address this need. ISE's *Aurora* towfish is owned by the Halifax-based Defence Research and Development Canada Atlantic, part of a 20-year-long working relationship between DRDC and ISE. It's also an integral part of the Canadian Navy's Interim Remote Minehunting and Disposal System (IRMDS). Other ISE vehicles built for government/military use include *Trailblazer*, a mine countermeasures ROV that has been put to work inspecting and recovering midwater and bottom mines by the military in the US, the UK, Norway, Denmark, Oman, Australia and Brazil, as well as Canada.

ISE also built two cable repair and recovery submersibles (CRS) for the US Navy's Military Sealift Command (the service providing cargo and troop transport for military use). These were used in the maintenance of the *SOSUS* (Sound Surveillance System) lines in the North Pacific and North Atlantic. *SOSUS* was a passive sonar system developed by the US Navy to track Soviet submarines. The sea acceptance trial for the ISE vehicles was held on the ship that had been designed to recover space shuttle booster rockets but was temporarily repurposed after the *Challenger* disaster. Don Muth recalls those trials vividly: "The weather was quite rough and the ship's freeboard was low, so we operated the system with 3 to 5 feet of water washing across the deck and around us. We were in harnesses, fastened to lifelines in order to keep from being swept overboard! However, the customer was pleased, as that proved that they had a system that could be operated in 'real world' conditions in the North Atlantic and Pacific."

ISE also built the *DSIS* (Deep Sea Intervention System) for the Canadian Navy to help with submarine rescue. The system was equipped with tools for deploying rescue pods, severing cable entanglements, and jetting to blow silt away from buried hatches, as well as hatch-opening tools. It was used extensively on recovery of the Swiss Air Flight 111 that crashed in St. Margarets Bay, Nova Scotia. ISE's *Trailblazer 25* was also used on that operation.

"On the morning of that crash, Jim was waiting in the parking lot for me to drive in at 6 a.m.," Don recalls. "Before I could even get out of my truck, he told me I had to get to Halifax that day to help in the inspection and recovery work at the crash site." Alas, flights were booked up at that time, so the first available seat was a red-eye leaving Vancouver just before midnight. Don arrived just in time to get the *Trailblazer* onto the ship and leave for the crash site. "We arrived late in the afternoon and started to dive immediately upon arrival. As we neared the bottom, I could see the scope of the tragedy. There were clothes floating by, as well as pieces of the aircraft, parts of suitcases. All this while we were still two miles from the point of impact!"

Don's first trip lasted two weeks before he returned home for a week. Then it was back to install the *DSIS* onto a second Navy ship to do more recovery, starting with the grim task of recovering human remains, a priority for family closure. That task continued for months, with various ISE employees working with the Navy. The job finally transitioned into recovery of the plane parts and cargo to assist in the crash investigation.

ISE's scientific track record continued with the completion of two *HYPER-DOLPHIN* ROVs in 1999 for Japan's Marine Science and Technological Institute, JAMSTEC. These vehicles allowed their scientists to conduct visual and filmed surveys at a maximum depth of 4,500 m (14,765 ft) using a compact, ultra-high-sensitivity high-definition TV camera.

The same year, France's National Institute for Ocean Science, IFREMER, contracted ISE to build all the components for the *Victor 6000*, an autonomous underwater vehicle (AUV) that operates without a tether. Designed specifically for scientific oceanographic work to a depth of 6,000 m (19,685 ft), this AUV is equipped with a high-performance navigation system and high-resolution photo and video imaging systems.

Over the years, ISE received government contracts for various local and global military projects on a contract-to-contract basis. And much of ISE's subsequent Arctic work with AUVs was driven by Canada's desire to establish its sovereign claim to the far North. Each of those missions required specialized vehicles and equipment that the company designed and provided. In 2011 ISE was named one of Canada's top 40 defence companies.

The AUV revolution

The *Victor 6000* was part of a new development that had been emerging for years. In the 1960s, subsea business evolved from divers to submersibles. That was followed in the '70s by a shift from submersibles to ROVs—when ISE first entered the picture. Certainly, the greatest advantage of an ROV is its ability to work underwater without endangering a human occupant. Instead, its human operator is safely on board a surface ship or platform, directing operations via commands carried through the ROV's tether. Alas, a major disadvantage of an ROV is that very same tether, or umbilical. The tether's primary function is to carry command signals and power down to the ROV so it can move and perform work; in turn, the same tether transmits data (especially visual) back up to the ship and the ROV's operator. But a tether also limits depth, manoeuvrability and access. It also adds weight, creates drag and carries the risk of entanglement.

Fortunately, many of the initial problems of getting data from ROVs were solved by using fibre-optic cable. As a result, today's ROVs can deliver high-definition video of remarkable quality. There's also an increasingly wide range of control functions, all without the risk of having human beings working at depth. But the newer side of the game is tetherless vehicles, commonly known as Autonomous Underwater Vehicles—AUVs.

For that evolution to occur, there were big questions to figure out. Was it possible to get rid of the tether? Could an autonomous vehicle be created that essentially has a "brain" on board? Could such a vehicle receive control commands some other way and carry its own power? The answers to such questions usually depend upon three factors:

- Technological advance: AUVs became possible when smaller and faster microprocessors became available, making the concept more practical.
- Mission necessity: Some missions, such as long under-ice surveys, were simply impossible for ROVs.
- Financial advantage: The cost of the surface vessels required to support some tethered operations was huge. Using tetherless vehicles could cut costs, sometimes significantly.

The evolution of AUVs was therefore the next logical development in the 1980s. As the "autonomous" part of the name suggests, AUVs operate independently of humans, using pre-programmed instructions. Those with no real-time contact with the surface can only have their data accessed once they are retrieved. Unlike ROVs, they have no physical tether connection with an operator, so essentially AUVs are self-powered and self-directed vehicles. The smallest are easily portable, weighing under 100 pounds. At the other end of the spectrum are more complex AUVs weighing thousands of pounds and requiring a dedicated support ship. Most AUVs carry their power on board, utilizing a variety of sources—specialized batteries, fuel cells or even a rechargeable solar power panel. Some, such as gliders, rely on gravity and buoyancy for propulsion. And the missions they can handle are increasingly complex and sophisticated.

With AUVs now taking a bigger bite of subsea work, the big deal is software, and these vehicles are getting even smarter. They're already good at seabed mapping and laying fibre-optic cable, but AUVs can locate and recognize objects of interest, such as submarines and explosive mines on the sea floor or in the water column. They can also monitor offshore resources such as fish stocks and track pollution from oil and chemical spills. Many AUVs can now be followed and interrogated from shore, ships, aircraft and even satellites, allowing for data exchange while still at sea.

One of the mandates for ISE's sister company ISER was to focus on designing the company's first autonomous vehicle. Before retiring from the Canadian Navy, James Fergusson had been a submarine commander, so logically he came up with the idea of a torpedo-like design for the AUV, using the same 21-in (53-cm) hull diameter as a standard Mark 46 torpedo. He asked Al Trice to do a preliminary design with the intension of using AUVs in the Arctic for under-ice survey work and subsea mapping. The federal government needed data in order to create reliable charts and construct a harbour for potential oil and gas tankers. But once Al studied the specs and the power source, he realized that the hull diameter had to be expanded to 27 in (68.5 cm) to accommodate the battery.

The company's first autonomous vehicle was the ARCS (Autonomous and Remote Controlled Submersible), developed in 1981 with production completion in 1983. However, after testing and sending a crew to the Arctic to check out operating conditions, the government's Arctic program was suddenly cancelled. ISE was paid for its work and allowed to market the vehicles, but it was a long dry spell with no sales and only a few small jobs. The company continued vehicle development, switching ARCS to lithium ion batteries as the technology updated. Then suddenly the company sold five AUVs in one year—two to France, one to Germany, one to Memorial University in Newfoundland, and one to the University of Southern Mississippi. Initially ARCS had not been designed for great depth, but the new clients wanted 3,000 m (9,840 ft) of operating depth, which required a redesign of the hull. Everybody was happy.

Then came a contract from the Canadian Department of National Defence (DND) as part of the joint US–Canada Spinnaker Project. It was for a much larger AUV that could lay long lengths of fibre-optic cable under the Arctic ice pack. The scope of this mission required a major re-thinking and re-design. To carry that amount of cable, the AUV would need a hull that was 50 in (127 cm) in diameter, as well as a lot more batteries, and variable ballast to compensate for the gradual loss of weight as the cable spooled out. That cable also had a minimum bend radius, so the design had to accommodate the cable being carried in the centre of the ballast tank. Given that it would be transported to the Arctic, the AUV also needed to be fabricated and shipped in sections that would fit into a light Twin Otter aircraft. As well, the contract specified that the AUV should be able to operate in both salt and fresh water. As with any military contract, it required volumes of paperwork.

Al Trice designed two compensating ballast systems for the new AUV, one for general vehicle stability and another variable ballast system for cable-laying.

Theseus, ISE's second major autonomous underwater vehicle, set an AUV endurance record of 60 hours under Arctic ice. Collection of ISE, Ltd.

Vince den Hertog worked on pressure hull design and other components. The result was *Theseus*, the company's second major AUV. Launched in 1995, it successfully completed two missions in the Arctic. The first laid a limited length of fibre cable to test and verify the system. The second of these stints, in 1996, laid several 220-km cables in 600-m (1,968-ft) water depths under an ice pack 2.5 m (9 ft) thick, establishing an AUV endurance record of over 60 hours under ice and accomplishing the full mission. *Theseus* is now available for charter from the DND. Its large payload bay can store and launch a number of smaller AUVs. And its variable ballast system enables it to be "parked" on the bottom for extended periods of time, until being called back to work by an external signal or a programmed predetermined time.

From September 1, 2017, ISE and ISER were amalgamated for business purposes and because many previously separate technologies were now overlapping, but the company's commitment to autonomous underwater vehicles has continued. Today, AUVs are a prominent business line for ISE. Its *Explorer* vehicles are the latest evolution from the company's earlier ARCS and *Theseus*. *Explorer* AUVs are modular, with each section about 3 ft (1 m) long. Essentially, they are tubes that can just click together, depending on the vehicle's intended mission. Some have a variable ballast system that fills one section. Some have some very expensive inertial navigation systems that continuously calculate position, orientation and velocity without any external references. That's important because once the AUV descends below the surface of the water, there is no GPS access anymore. So it's critical to have equipment on board that keeps track of where the vehicle is. And, of course, there are all the sonars, sub-bottom profilers, multi-beam sonars and salinity meters that get packed into those tubes, depending on the vehicle's mission. *Explorer* AUVs are versatile and can be tailored to do just about anything required in the ocean, whether it is exploration, mapping the underside of icebergs or even laying cables.

Today, the Japanese Coast Guard utilizes two *Explorer* AUVs for mapping the seabed in order to update their marine charts in various parts of the world. Another *Explorer*-class AUV was delivered to the Centre for Marine Environmental Sciences (MARUM) at Germany's University of Bremen. Named *SEAL*, this AUV has a working depth of 5,000 m (16,405 ft), an endurance of 24 hours at a speed of 1.5 m/s (3.35 mph), and a payload capability of 100 kg (220 lbs). Recently, the China Ocean Mineral Resource R&D Association (COMRA) ordered ISE's first 6,000-m (19,685-ft) AUV so they can work deeper.

Feedback—the learning goes both ways

In 2018, ISE delivered a 5,000-m (16,405 ft) *Explorer* AUV to the University of Tasmania (UTAS) and the Australian Maritime College (AMC). Designed for Antarctic operations under the ice shelf, it's loaded with an on-board variable ballast system that can be adjusted while running under the ice, a sub-bottom profiler, a sonar and side-scan sonar with an option for bathymetry, and a CTD (an instrument that measures the conductivity, temperature and density of water).

A year later, ISE expanded its expertise and experience working with AUVs in icy conditions during a three-month test in Antarctica for the University of Tasmania. In the process, ISE became the first company to design, produce and test AUVs in both Arctic and Antarctic waters. Jean-Marc Laframboise is a senior technical adviser at ISE, as is colleague Don Muth. As part of his job, Jean-Marc accompanied the *Explorer* AUV, which the University of Tasmania named *Theseus*, to Antarctica.

In doing so, Jean-Marc spent 93 days with UTAS personnel and the Australian Antarctic Division (AAD), part of the government ministry that manages about a third of Antarctica. But it was not a straightforward journey.

A heavily bearded Jean-Marc Laframboise (left) from ISE with the University of Tasmania (UTAS) crew and their AUV *Theseus*. Personal collection of Jean-Marc Laframboise and ISE Ltd.

Besides the extensive preparation with UTAS, there was a lengthy selection procedure. "Getting accepted to work there is no small feat," he explains, "because once you're there, you're there for your entire stint. From a family point of view, it's a big commitment. When you say goodbye, it's goodbye for a long time.

"The selection procedure for Antarctica begins with a thorough physical exam. After that there's induction training modules for everything—safety on ice, first aid, working in the boats, life on the station, the equipment, the weather, penguins' approach-safe zone, and survival on the ice. You even learn about 'human waste management' as it's now called, because the rule is 'garbage in, garbage out.' I had been to the Arctic many times so had some experience, but it's different with AAD. Everybody must take the courses. And there are rules for everything, but a lot of those are based on lessons learned the hard way, so it's good."

In Antarctica, or on any job like this, weather determines everything. When Jean-Marc and other workers and scientists arrived by ship at Davis

Station, the shore and the bay were covered with ice, so there was a three-week wait before getting the AUV in the water. Its mission was to go underneath the Sørsdal Glacier Tongue, the part of the Sørsdal Glacier that projects into the sea. If possible, it was to measure water temperature and get sonar data in order to create a profile underneath the ice. At the time, no one knew what the glacier looked like underwater. The glacial tongue rose about 60 m (200 ft) above the waterline, and the guess was that it continued about 600–700 m (2,000–2,300 ft) below. "The trick to putting an expensive vehicle like the *Explorer* in the water here is to remember that apart from the glacier there are other obstacles, as well as very shallow water. Plus, there was no chart to provide any information." Getting to the glacier required surface transit, but once at the Sørsdal Glacial Tongue, *Explorer* dove underneath and mapped it. "That's how we discovered that the glacier was actually 1,200 m deep!"

"The AUV did very well, performed its missions and came back," Jean-Marc reports. "As we say in the AUV world, if the number of launches equals the number of recoveries, then it's a good day! My job going to Antarctica was to help the UTAS AUV team with the big picture and to learn the risk factors of working in that kind of environment. Of course, we did a lot of prep ahead of time, working with team leader Peter King and the engineers from the Australian Maritime College."

Once Jean-Marc got back, there was an equally important engineering debrief with ISE staff, discussing the challenges, what wasn't 100 per cent on the vehicle and little tweaks that need to be addressed for next time. "It's this hands-on learning experience that makes what we do here so realistic and so valuable."

Company history and core employees

During the early years, ISE was in a small Port Coquitlam workshop. Then the company moved to Port Moody, BC, and successfully built up the business. Eventually the ISE group of companies bought land and built their present complex back in Port Coquitlam, which opened in 1985. That base also provides close proximity to the protected waters of Indian Arm, which means the company can test vehicles 365 days a year and never get blown out because of weather. It also requires only a short drive to get aboard the *MV Researcher*, which ISE launched in 1985 as a purpose-built trials vessel and testing platform for their undersea vehicles. The far end of Indian Arm offers a 200-m (660-ft) depth for test dives; if more depth is needed, there's Jervis Inlet, the deepest fjord on the BC coast, with a maximum depth of 732 m (2,402 ft). For a vehicle rated to 2,000 m (6,560 ft), that's a pretty effective testing depth, Jim maintains. "Besides, if it's going to leak, it's probably going to leak in the first 10 minutes."

Like the majority of the province's subsea businesses, much of the company's work is little known or understood by locals, although in 2014 the Tri-Cities Chamber of Commerce chose ISE as its Business of the Year. Wisely, over the years, Jim McFarlane reached out to a number of local, provincial and national government officials, beginning with John Manley when he was Liberal MP for Ottawa South. The visits of other notable dignitaries, including BC Premier Gordon Campbell and several astronauts, are all pictured in photos on the company walls. And certainly Jim himself has been significantly recognized and commended: he's an Officer of the Order of Canada and has a long, impressive list of awards from the subsea industry, as well as four honorary doctorates. That said, the need for a company to connect continually with local and national figures is crucial, ensuring they realize that this high-tech capability exists in Canada. It's also what's required to keep any business moving forward.

From the beginning, ISE and Jim McFarlane attracted an amazing array of talent. A tiny sampling includes:

- **Mike Macdonald** brought his HYCO expertise to ISE in 1977, becoming the company's first executive vice-president. He piloted ROVs around the globe, set up ISE's Houston office, served as sales engineer and then started training Vickers's pilots. He retired as "chief pilot" and left ISE in 2010.
- Subsea pioneer **Al Trice** joined the company a few years after HYCO folded. It was his first introduction to autonomous underwater vehicles, and he became project manager, heading up research and development of AUVs for ISE's sister company ISER.

- **Eric Jackson**, now head of Cellula Robotics, started at ISE right out of university in 1978 and stayed for 22 years. "Back then I was probably the only electrical engineer in the company so I became chief control systems engineer and managed projects ranging from force-reflecting robotic manipulators to AUVs such as ARCS, and ISE's Canadian Space Agency projects."
- ISE also hired **James Fergusson**, the youngest sub captain in the Canadian Navy at the time he retired from active duty in 1981. He became the head of International Submarine Engineering Research (ISER), a career that absorbed and captivated him even after retiring a second time in 2013 and becoming a consultant.

Then, as now, the heart of International Submarine Engineering is its employees, their diverse experience and expertise, and their commitment. They range from full-time specialists and behind-the-scenes technicians to those brought in to work on various projects. There are many specialists on the company roster—systems engineers, control engineers, mechanical engineers, software engineers and electrical engineers. There is a similar range of technicians. And plenty of software whizzes. A few employees hired out of high school have amassed their own wealth of capability and experience. And let's not forget the various dogs that comfortably roam from office to office.

Typically the company attracts co-op students from various Canadian technical institutions and universities. Once they've put in their prescribed work time, they often go back to school to finish their degree and then return to work their entire career at ISE. Many of ISE's staff have 30 to 35 years of experience; "junior employees" usually have been with the company for a decade at least. Since much of the job is product-specific, the company often recruits new hires right out of local schools and teaches them how to do what ISE does, what the product is and how it works.

There's a saying at ISE that "prior planning prevents piss-poor performance." That is backed up by the company's breadth of experience in the field. Walking the halls with James Sr., we pass a conference room in which two of the five people have been to the North Pole, one to both poles. There's also an impressive photo gallery of underwater vehicles with names like HYDRA,

In the Arctic, ISE's AUVs recharge their batteries, exchange data and upload new instructions without ever leaving the water. Collection of ISE Ltd.

WRANGLER, SUPERDART, DYSUB, ARCS, DOLPHIN, and RASCL. Each photo has a story, often a subsea discovery, and features missions around the globe—India, Korea, Africa, Australia, Germany, Japan, China, France, Newfoundland, and under Arctic or Antarctic ice.

As he strolls down the hall, Jim introduces Steve Nishio, an Arctic veteran who describes a recent trip: "Chris Kaminski, Kevin Young and I were up in the North Pole two years ago. Our main camp was the Canadian Hydrographic Service; but we deployed the AUV from a remote camp where we stayed and worked in insulated tent structures. While on the mission, the vehicle was able to stay in the water the entire time, because we used a special system we called a 'catchie' that bored through the ice and clamped onto the vehicle, charged it and exchanged data. That AUV travelled more than a thousand kilometres without ever coming out of the water!"

Steve also explains that everybody participating in a remote mission has many different areas of expertise. "That lets us cut down on crew or lets us swap somebody in for somebody else. It's always an interesting challenge."

What is obvious is that whether employees are in the shop, out on the water or under polar ice, it's all in a day's work at ISE. And the numbers don't lie. A recent tally totalled more than 400 robotic manipulators and 240 vehicles that ISE has built and delivered in its 46 years of existence.

We changed the world

Years ago, Jim McFarlane put together a small private booklet with the cheeky title *We Changed the World*. "And we sure as hell did!" he states with pride, though he qualifies the title by saying it refers to changing the subsea world.

In the first pages he details the events of the 1960s that fostered a period of technological revolution:

- Jacques Piccard and Don Walsh became the first humans to reach the Challenger Deep in the Mariana Trench (the deepest part of the ocean) aboard the DSV *Trieste*.
- The *Concorde* flew and men landed on the moon.
- Mixed-gas diving was investigated by Jacques Cousteau (France), Hannes Keller (Switzerland), Capt. George Bond (USN), as well as Phil Nuytten (Can-Dive), Lad Handleman (Cal Dive) and Comex.

In the book, Jim briefly details how subsea equipment developed in the late 1960s, '70s and '80s, and how system operations rapidly moved ahead in British Columbia to include undersea work capability in manned subs, ROVs, AUVs and hybrids. Subsequent pages are a veritable *Who's Who* of the subsea industry in BC: Phil Nuytten with Can-Dive and Oceaneering; HYCO's Don Sorte, Mack Thomson, Al Trice and Mike Macdonald; sonar's Willy Wilhelmsen, Helmut Lanziner and Jim Kail; investor John Horton; Atlantis tourist submarine's Dennis Hurd and John

Jim McFarlane sports an ISE baseball cap emblazoned with a typical company motto that speaks to the accomplishments of his illustrious career—Git 'r Done! Photo by Harry Bohm

This photo taken by Al Robinson captures fellow BC subsea luminaries seated at the Pioneers' Table to celebrate Jim McFarlane's Lifetime Achievement Award from the Historical Diving Society. Left to right: Helmut Lanzinger, Willie Wilhelmsen, Jim McFarlane, Phil Nuytten, Al Trice, Mike Macdonald and John Witney. Personal collection of Al Robinson

Witney; David Fissel in ASL Environmental Sciences; and Al Robinson of Inuktun.

Jim McFarlane sums it all up: "In some respects, it was serendipitous that we had the facilities here in British Columbia to make stuff that could be tested and worked on that enabled us to go out and change that world. It was the right place at the right time with the right equipment and the right science." Amen to that.

CHAPTER 9

Sonar: The magic of seeing underwater

Mid-1960s-present day

———

Mark Atherton's fascination with underwater imaging and technology started in 1976 when he got a job with Phil Nuytten's Can-Dive. Mark had worked in underwater photography and often assisted Helmut Lanziner, who ran Can-Dive's Offshore Survey Division. "Side-scan sonar was still new to Canada," Helmut recalls. "I was the only one using it in those early days, so I talked with Phil about needing to train more side-scan guys as well as positioning operators." Phil agreed, knowing that there was no shortage of work or money to be made, particularly in the oil industry in the Arctic. Helmut had first worked with Garry Kozak, now considered one of the foremost sonar experts of North America. So Helmut took that teaching experience and set about organizing a realistic training session for several apprentices. Mark Atherton was one of them.

Helmut wanted his students to find a sunken tugboat called the *Gulf Master*. The tug had gone down off the Sunshine Coast near Sechelt, BC, in 1967, with the loss of five crew members. Earlier searches using a ship's sounder and magnetometer had been fruitless. The wreck of the *Gulf Master* still hadn't been located—a perfect challenge for Helmut's students.

Armed with an early prototype model of an EG&G side-scan sonar for the search, along with microwave positioning equipment, Helmut and his trainees headed out to Sechelt. Helmut explained to his students the importance of advance homework. He had already interviewed three shore-side witnesses, getting the usual skewed accounts. He had also obtained a suggested triangular "area of probability" from Transport Canada. It was an area Helmut knew, having passed it on tugs and other vessels. To him, that data didn't seem to make sense, so he widened the search area slightly. Helmut had also established a microwave station in a secluded area at the top of a cliff at the University of British Columbia that offered a visual line of sight to Sechelt. Then he directed his students to set up other microwave positioning stations ashore near Sechelt.

"The other important part of sonar success, in addition to the equipment itself, is having an exact reference point for your location," he explained to them. "That ensures that you travel straight lines when towing the sonar and don't miss anything. I always made the choice never to go on a search job without accurate positioning. You have to be 100 per cent confident that the area you've searched can then be ruled out. Otherwise, it's a waste of time."

Finally, it was time for the students to put the side-scan sonar in the water and begin a search grid for the new area *beyond* the original suggested search triangle. Within a couple of hours, they had found the tug! Mark Atherton recalls that discovery vividly.

"The resolution of the side-scan wet-paper recorder was about eight shades of reddish colour that varied depending on the signal levels received by the side-scan transducer. I distinctly remember watching this blob appear on the sonar record. When scaled, it was about the right length of the tug we were trying to locate. And there was a shadow behind the blob indicating that the wreck was on, but proud of, the seabed. That was the first shipwreck I ever saw on side-scan. Looking at that blurred image, I remember thinking, 'This is the most incredible instrument I've ever seen.' And then I realized, 'This is what I've gotta' do.'"

Sonar basics

As a kid, I dreamed of a pair of magic glasses that would let me see underwater. But what I never imagined is that some attempts to "see" underwater would use sound rather than sight. As I got older, I learned that dolphins, bats and some whales use sound to locate food and avoid enemies or obstacles. Once I understood how that worked, I had a very basic understanding of how sonar functions. Of course, things quickly get very technical. So let's begin with the basic vocabulary of the sonar world.

Lesson #1: A sonar transmission sends out a pulsed sound wave signal, or **ping**, from a **transducer**. If this acoustic energy hits a **target**, whether stable or moving, it gets returned back as an **echo**. A series of these returned echoes, shown closely spaced on the display screen, roughly outlines what's in the water. A **sounder** is simply another term for a basic acoustic beam going out from the bottom of a boat or ship.

Lesson #2: Generally, there are two basic types or applications of sonar— **imaging sonar** and **profiling sonar**. The significant difference between the technique of these two types of sonar is the shape or pattern of the beam of sound pulses that's being sent from the transducer.

The return echo of an **imaging sonar** gives a general picture of all that's on the bottom, much like a photo does (see Figure 1 and 2).

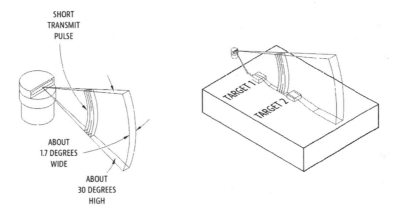

Figure 1: Typical fan-shaped beam as used on imaging sonars

Figure 2: Fan-shaped sonar beam intersects with a flat bottom and targets

An imaging sonar uses a fan-shaped beam that scans an area in one of two ways:

1. It can be pulled in a straight line behind the vessel (mounted on a towfish), in which case it's sending pulses out to the side, so is called a **side-scan sonar** (see Figure 3).
2. Or this-fan-shaped beam can be mechanically rotated (see Figure 4).

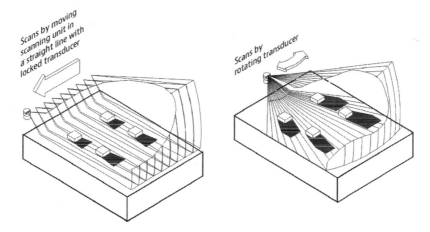

Scans by moving scanning unit in a straight line with locked transducer

Scans by rotating transducer

Figure 3. Side scan sonar

Figure 4. Rotary scan sonar

A **profiling sonar** uses a conical beam and can take more time because it's shooting each object with individual acoustic pulses (see Figure 5). The result is a series of digital returns that can tell exact dimensions and distances rather than a general picture. For example, a profiling sonar can determine exactly how deep the underside of an ice ridge is (see Figure 6).

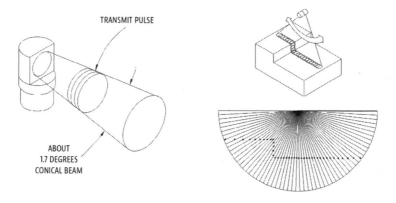

TRANSMIT PULSE

ABOUT
1.7 DEGREES
CONICAL BEAM

Figure 5. Typical pencil-shaped beam as used on profiling sonars

Figure 6. Plot digitized echo returns to show profile (cross section) of bottom

Not surprisingly, many sonar jobs utilize both applications to provide a general image as well as the actual physical relationship data. Given good water conditions and an experienced sonar operator, the resulting images either enable identification of the chosen target or produce accurate measurements.

Sonar applications—imaging and profiling—were soon adopted and adapted for other subsea applications as the need arose. For example, in order to lay communication cables, a cable ship needed a detailed picture of the ocean floor. Or to position an offshore oil rig, oil and gas companies needed to know what was on the bottom. And if that oil rig was in ice-covered waters, as many were, the company needed to determine exact measurements in order to dredge depressions deep enough to protect their wellheads from the ice that could reach far below the surface.

Ships or boats commonly deploy sonar, but ROVs now use sonar, too. Initially, ROVs had to travel underwater to locate a target. But an ROV equipped with imaging sonar can also sit stationary on the bottom while the sonar scans the surrounding area. ROVs and AUVs both depend on sonar for long-range obstacle detection and avoidance.

Today, there's an amazing variety and combination of these basic sonar technologies—all are essentially a high-tech version of the magic glasses I imagined as a child. These developments are the result of pioneers such as electrical engineer Harold "Doc" Edgerton at the Massachusetts Institute of Technology, who

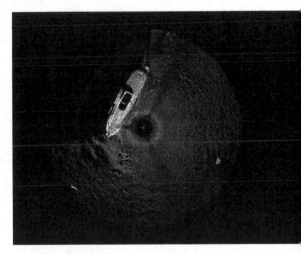

This high-resolution-imaging sonar scan shows remarkable clarity of the wreck of the *Condor*, a blockage runner for the Confederate army during the American Civil War. She sank during a hurricane just off the coast of North Carolina on October 1, 1864. Rose Greenhow, a spy for the Confederates, died in the sinking; her body washed up on shore and her dress contained jewelry and documents. Image by Brian Abbott, courtesy of Mark Atherton

along with his student Martin "Marty" Klein pioneered the development of side-scan sonar in the 1950s. In the mid-60s, Willy Wilhelmsen and Helmut Lanziner became Canada's sonar pioneers, leading the way for others.

Willy Wilhelmsen and Helmut Lanziner

Willy came to Canada in 1962 from Kirkenes, a small Norwegian village only 5 km (3 mi) from the Russian border. "Working in the Arctic later on was nothing for me because I grew up in those kinds of conditions. As a kid I'd walk a kilometre to get milk, and by the time I got back home, the milk was frozen." By age 11, he was already interested in electronics and remembers drawing circuit diagrams in the snow.

Emigrating to Canada as a young man, Willy went to work for George Kelk & Co., then moved to a small private company in Toronto with the impressive-sounding name of Canadian Research Institute Ltd. The company employed technically educated immigrants who were recent arrivals. Many spoke German, with only limited English. While building voltmeters and measuring devices, Willy met Helmut Lanziner. He recalls that Helmut was the only one in the company who spoke English most of the time. Helmut had been born in Munich, Germany, and came to Canada with his parents in 1957, at the age of 17.

In the early 1960s Willy moved to British Columbia, working with Nova Scotian Stan Davis at GTE Lenkurt Electric in Burnaby. "I got the highest-paid job as a technician, so I was hated for a couple of days, but then things settled down," he chuckles. "That's also where I met my wife, Deanne. She was working there too."

In 1966, Stan Davis suggested that he and Willy start a company of their own. They named their venture Aqua Electronics, an accurate description of their intended scope of work. They began by building electromagnetic underwater communication systems, essentially an underwater telephone for divers, and Helmut soon joined in. Willy explains, "An underwater telephone uses basically the same electronic components as you'd find in a walkie-talkie. The main difference is that radio frequencies use electromagnetic signals with an antenna. In an underwater telephone, the antenna is replaced with transducers, which use sound waves. When in transmit mode, the transducers transform electrical signals into pressure waves. In receive mode, the reverse is true—the same transducer converts pressure waves to electrical signals. For a diver telephone, of course, the microphone and the ear phones have to be waterproof. Also, a mouthpiece and full-face mask are necessary so that the diver can talk. Underwater telephones use frequencies from 10 to 30 kilohertz, which gives a range of a couple of kilometres to 10 kilometres. By comparison, radio frequencies can provide communication for ranges up to 10 to 20 miles."

"The big stumbling block," Willy adds, "was a connector that would hook up the telephone's underwater microphone to the mouthpiece. We couldn't afford to buy the expensive underwater mateable connector, so one night, it was Helmut and I and a bottle of vodka ..." Helmut laughs and adds, "hardly ever!"

"We designed one using a transformer," Willy continues, "and it's still in use today. The principle of this type of connector is also used for downloading data from AUVs." When Aqua Electronics ended, Willy bought the equipment and started up Subcom Systems in 1968. Helmut became a partner in the new business, along with divers Jim Kail and Matt Matthews.

Helmut recalls how their tiny Canadian company sold systems to the US Navy: "We went to a demonstration and were competing against the top US military contractors and engineers. They showed up in their big transport trailers and were getting their equipment ready, when the two of us pulled up in our car and just opened the trunk. They thought we were a joke. A guy from the company parked next to us came over and asked, 'Who *are* you guys?' We told him we were from Canada. Then he asked, 'Well, who's financing you?' And I said, 'We are.' They didn't believe us because everybody at these demos was a major player. Ironically, we were the ones who won the contracts after our demonstration. We had a diver swim out until we could barely see his bubbles, but we were still communicating with him! The military guys actually turned out to be helpful and would say, 'Have you talked to so-and-so? You should go see them because they're looking for this kind of equipment.'"

A year or so after Subcom started, Helmut left to work at Can-Dive. Subcom Systems continued until 1972. "In the end, the assets were taken over by Tom Murdock of Orcatron," Willy explains. "Technically, the company was quite successful. We sold a lot of stuff to the Canadian and US Navies. I remember a single order for 30 systems. Our communication equipment was for divers, and we were the only ones making that."

In 1972 Willy started Mesotech Systems with three partners: machinist Erling Kristensen, mechanical engineer Alan Mulvenna, and electrical engineer Bob Asplin. "We were busy doing underwater telephones, transponder systems, positioning systems and pingers. Fortunately, we got a big contract with Dr. Igor Mikhaltsev and Dr. Anatoly 'Tolya' Sagalevich from the Soviet Union when they were over here for their *Pisces* submersibles. We ended up with a $250,000 contract for all the electronics and spent a year producing 200 pieces. They couldn't buy anything from the Americans, so we had to build everything—underwater telephones, high-powered positioning systems for submarines, transponder systems, pingers, etc. It was all done above board with permits. That Russian contract got us going!

"Mostly I remember going to Deep Cove on a cold November day for the final acceptance test of the equipment. It was Tolya Sagalevich, Bob Asplin and myself. The day before, Tolya had said, 'Bring something warm.' I said, 'Yeah, lots of coffee.' But he had other ideas. 'No, no. Not coffee.' So the captain of the surface landing craft we were using came back with two bottles of vodka and whiskey and we loaded it aboard, along with the equipment. We left Bob on the dock, shivering, while Tolya and I took off in the boat, going out a mile with the hydrophone in the water. 'Can you hear me, Bob?' 'Yeah, loud and clear.' With that, Anatoly said, 'The customer is happy. Bring out the drinks.' So the drinks were served and the day was lost after that."

Mesotech employee Gordon Kristensen recalls, "Mesotech was doing things that had never been done before, things that Helmut needed when he was working in the Arctic. Willy would design the equipment, we'd build it, and Helmut would put it to work. The positioning that Helmut did in the Arctic was all new."

Willy's daughter Disa has fond memories of growing up with Mesotech. When she was less than five years old, she tagged along to the office with her dad and learned how to use side cutters. Later one of the ladies at Mesotech taught her how to use a soldering iron. When she was seven, Jeff Patterson, an electronics assembly technician for Mesotech, introduced her to a heat gun. Not surprisingly, for her grade 6 science fair project, Disa chose sonar. One judge gave her a C– because he thought she couldn't possibly know what she was talking about. But she did!

All five Mesotech partners, shown on the day Mesotech was sold. Standing: Bob Asplin, Erling Kristensen and Willy Wilhelmsen. Seated: Alan Mulvenna and John Horton. Personal collection of Willy Wilhelmsen

At that time, Mesotech was in the same building as John Horton and Horton Maritime Exploration (HME). Horton became a shareholder in Mesotech, investing $10,000 at a point when Mesotech needed the capital, which got him 75 per cent of the company. However, as Mesotech grew and the four owners could afford it, they bought back shares and ended up with 50 per cent. Of course, John Horton still had 50 percent, too, so when Mesotech sold in 1985, he received two million dollars for his investment!

Helmut and Can-Dive

In 1970 Helmut went to work for Phil Nuytten and Can-Dive, running his own separate division with side-scan sonar and underwater cameras. Helmut recalls, "Side-scan sonar was a new thing in Canada, and most people didn't know what you were even talking about. I had heard about what a powerful tool side-scan was, but I didn't really understand a lot of it. So Can-Dive brought one in, and I started playing around with it, learning on the job. I was perhaps the first person in Canada to be doing that. Within a year I realized that side-scan sonar was this exciting new imaging tool for finding shipwrecks and all kinds of lost things that previously seemed impossible to locate."

Helmut spent years working in Canada's Beaufort Sea, first for Can-Dive and later for his own company. "When I arrived in the Beaufort, people thought I was some sort of magician because not only was I finding things, but I was doing it in a shorter time! I wasn't a magician; I was just using the new sonar tools. Mesotech's great side-scan sonar and also accurate positioning made a day-and-night difference. You went from spending years trying to find a shipwreck to locating it in a few hours—a totally different ball game!"

Sonar would prove particularly useful with the offshore drilling industry in the Beaufort Sea and other places in the High Arctic, as well as on both the east and west coasts of Canada. Dome Petroleum was just getting started, and Helmut was doing most of the surveying work for Dome via Can-Dive contracts. "Most of what I was doing in the Arctic was usually with Mesotech equipment, so Willy and I talked a lot. I provided feedback on the equipment, how it was being used and the way it was working."

After years of underwater surveying for Can-Dive, in 1977 Helmut created his own company. It started as Offshore Survey and Positioning, but its name was quickly shortened to Offshore Systems Ltd. (OSL). "Willy offered me space in his place for my new company, which made sense because I was usually doing the offshore surveying with Mesotech equipment."

In the 1960s, oil and gas companies began drilling further offshore and deeper in the Gulf of Mexico; in the '70s, they moved to the North, where new oilfields were being discovered. Drilling efforts now had to contend with ice in various forms: icebergs, pressure ridges, thick pack ice covering the water, or, in some places, a short season of ice-free water.

No matter where the drilling occurred, companies desperately needed to "see" the bottom in order to find the most suitable place to position a wellhead. Sonar imaging was the solution. "As early as 1974," Helmut recalls, "we started trying to get sonar data in the North, mostly working from drill ships and going down through the ship's central moon pool. Or if we were in the High Arctic, then we had to gather sonar data by drilling holes in the ice and lowering the equipment."

The Western Arctic and the Beaufort Sea don't have icebergs; instead there are pressure ridges. That's where two pieces of pack ice slam together forcing ridges up, down, or both. The upside-down pressure ridges were the main concern for oil companies, since the ice can reach all the way down to the sea floor, even in relatively deep water, leaving scars or tracks on the seabed that are visible on some side-scan records collected in the 1970s. Like icebergs, these pressure ridges presented the potential for losing an oil well in the Beaufort.

So when working in the Western Arctic, companies like Dome Petroleum and Canmar initially dredged what were called "glory holes" in the soft bottom sediment in order to position a wellhead below the mud line on the bottom. The thinking was that if the hole was deep enough, a

This photo shows Mesotech's first Arctic deployment of its revolutionary pipe-trench profiling sonar, Model 952, in 1977. Helmut Lanziner (at left) is just about to lower the sonar through the rotary table on the drill floor, and below that through the ship's moon pool (a rectangular-shaped hole through the rig's main deck) to conduct a profile/search of the seabed. This photo was taken on a semi-submersible drilling rig in the Beaufort Sea (Canadian Arctic). A pipe-trench profiling sonar is now described in Imagenex's catalogue as a "Mechanical Scanning" sonar because the scanning part of the sonar head is motor driven, so it rotates mechanically. Personal collection of Helmut Lanziner

passing ice sheet wouldn't tear the wellhead off. As these large holes were being dredged, the challenge was to determine whether the holes were symmetrical and deep enough. To complicate things, the measurements had to be taken from a fixed position off to the side, well out of the way of the big dredging machinery cutting the glory hole.

Sonar was the obvious solution to define the exact dimensions—the shape, size and depth—of the dredged area, but such equipment didn't exist. Willy recalls Terry Thompson coming by Mesotech and explaining the need for some type of sonar that could generate a cross-sectional profile of a

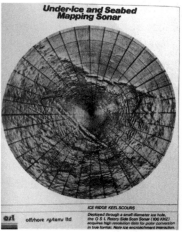

Willy describes Helmut's adaption of side-scan to work in the Arctic as "a brilliant piece of equipment." In 2020 the two demonstrated how the cylindrical mirror cone could translate rotary data into a linear profile of the bottom, sometimes even showing the scour line of a deep pressure ice ridge. Photo by Vickie Jensen. Photo of sonar record is from the personal collection of Helmut Lanziner

glory hole to show how deep or symmetrical it was. He showed Willy an old Japanese manual for a similar type of sonar that was no longer in production, asking, "Can you make one of these?" Willy went to work and three months later had a prototype of a profiling sonar for Helmut to test in the Western Arctic. This would become Mesotech's revolutionary pipe-trench profiling sonar.

Many drilling jobs presented new sonar challenges. For example, the Drake Flowline Project of 1978 was the first pipeline constructed *under* the ice in the Canadian Arctic. It was a demonstration project to see if a pipeline could be laid at significant depth, and in waters covered by thick ice much of the year, in order to connect the wellhead with onshore facilities. The project required sonar records of the seabed to see if there were rocks or boulders that might impede dragging the pipeline to the site, so Helmut and his company's team were given that task. But the ice cover on the water immediately presented challenges.

"We were using a standard side-scan sonar to start with," Helmut explains. "Normally, it gets pulled through the water in a linear fashion so the transducer can fire out each side and provide data that gets recorded. But when you're trying to survey through holes drilled in the ice, it isn't possible to tow side-scan in the traditional manner. Instead we tried putting the sonar down through the drill ship's moon pool and then rotating it in the water. Every 10 degrees, we'd mark the side-scan records.

"That gave us a record displaying linear information fired out to the side, only it was distorted and circular. But at least we had range and bearing, which was what was needed in order to detect the wellhead. However, once the oil companies started asking us to do mapping of the seabed in the ice, we had to figure out some way to de-skew all this data. Finally, I came up with a different idea. It involved a cylindrical mirror cone in the middle, surrounded by the data record. By looking downward at it, we could convert all the sideways sonar data back into reality."

Helmut recalls that the real test came in 1978 during a pipeline survey across McClure Strait, on the edge of the Northwest Territories. "It was a really big job across 80 miles of ice. Mark Atherton recalls the genius of Helmut Lanziner's conical adaption for side-scan: "It was a vertical pole that had a side-scan sonar inside. We cut a hole every 200 metres and then the pole was lowered through each hole in the ice. The sonar was then mechanically aligned horizontally and the pole turned using an electric motor and chain-drive arrangement.

"To determine orientation, Helmut used a huge protractor ring that had the degrees of the compass machined on it. The operator would start the motor, which slowly turned the pole/sonar in a circle. This generated a polar image on the side-scan recorder's traditional display, printing line by line as it created a record. After a couple of the motor turns in the icy hole, the system was stopped and the record annotated; then the operator flipped the side-scan up into the pole, extracted the pole through the ice, and moved to the next augured hole."

"It worked a whole lot better than I thought," Helmut admits. Then he laughs and adds, "I used to talk to people about my 'converter' for this data ... I didn't say it was just a mirror. In fact, it *was* very tricky getting rid of the reflections and putting the data together optically. The record is only a few hundred metres long, but it's all mosaiced together, and you can even see all the ice scour lines on the bottom. I don't know if anybody else has done that. I just did it out of necessity."

Mark Atherton recalls that oil service companies were so impressed by the sonar's performance that everybody wanted one. The name Mesotech became synonymous with this new sonar, much like Xerox is to a photocopier or Kleenex is to tissue. "When an offshore call came in asking for a 'Mesotech survey,' everyone knew what equipment was being requested." By 1983, this new technology established Mesotech as a leading sonar supplier to the offshore industry.

Later on, drilling companies like Panarctic Oil came up with a different technique to situate their drill rigs in the High Arctic, one that was uniquely Canadian. Helmut explains that during the winter the company would create a large ice island on which they would eventually build an entire drill rig and camp. They began by pumping tons of ice water onto the surface of the pack ice, freezing each layer before repeating the procedure to gradually build up the surface. When the area was big enough and thick enough, a Hercules transport would land and drop off living quarters for a crew, as well as food and equipment. Then pumping continued, thickening and lengthening the landing strip until Boeing 727s could bring in entire drill rigs piece by piece.

"It was almost unbelievable," Helmut states. "They could set up an entire drill rig in 10 days where there had been nothing before.

"One of the problems of these ice platforms was movement. If the whole thing began to move laterally, the drill rig could only tolerate so many degrees of movement before having to pull the drill pipe. So Willy developed an ice-motion monitoring system with audible warnings. We'd install that when the rig was being readied for drilling. And we also set up underwater TV monitoring systems on all drill rigs. As well, we were also flown out by helicopter to set up microwave transponder systems on the ice. Regularly, we had to deal with polar bears. All this, in addition to sonar work!"

The challenges of Arctic work

Anyone who has worked in icy waters and heavy seas or tropical heat and humidity, knows that equipment that functions perfectly in moderate scenarios can suddenly get cranky and malfunction in extreme conditions. Engineers back in their warm office buildings often haven't anticipated those problems, nor do they know how to quickly solve them. That knowledge only comes from working the equipment in those conditions. Willy states in his matter-of-fact way: "People talk about us as crazy risktakers, but at the time we didn't know we were taking risks. We were just trying to make things work."

If extreme conditions challenged equipment, the same is true for humans. Helmut vividly recalls working in -40° temperatures. Personal collection of Helmut Lanziner

Helmut recalls a series of contracts with the US Navy working in the Chukchi Sea between Siberia and northern Alaska. Initially the job involved "submarine business," which meant the Navy couldn't use non-US citizens without a long clearance process. "Since there wasn't time for that, we suggested that they send a team of acoustic experts to Canada and we'd train them.

"They did that, but in the end, the new team only got about 17 per cent of the data we could have got. The problem wasn't with the sonar equipment or their operation of it; it was with the ice. They just didn't know what to watch for with the deep cold. For instance, if we weren't going to be able to continue working with the equipment, we knew not to pull it out of the water and lay it on the ice. Instead we would put it back in the hole because the water temperature was only going to be 0°, −1° or −2° Celsius. When they left equipment out on the deck of the ship or on the ice, things wouldn't work because the air temperature was so much colder than the water temperature. Also, different rubbers and compounds behave differently in those extremely cold temperatures. You only learn all that by working out there and by equipment failing. We just had lots of experience to draw on and they didn't. Eventually we got clearance and did the work."

Helmut was acutely aware that getting an accurate geographical position in the Arctic was also extremely difficult, given that GPS was not yet operational. Fortunately, sonar advances were becoming more common. So was the use of microwave transponders, acoustic underwater transponders and

ROVs—all opened new possibilities for more accurate data. Initially, Helmut's company OSL suggested that ships operating in shallow, icy conditions have vessel-specific data systems installed. These showed a chart background and numerically detailed the ship's relationship to the channel centre and shoals. OSL created and maintained this electronic chart data—and it worked! The system gained fame and spread to other locations and fleets. Over the next several decades, as computer and navigation systems improved and radar overlay was added, these were integrated into Helmut's system. In 2005, Helmut was awarded the Order of Canada for his work in navigation and charting systems.

The start of Imagenex

In the mid 1980s, Mesotech was building transponder systems and pingers for Honeywell, Simrad and other competitors. As time went on, Simrad bought more and more, and in 1985 the company made an offer to buy Mesotech. Willy was not keen to sell, but his partners were. (A decade later, Simrad Mesotech was acquired by Kongsberg, becoming Kongsberg Simrad Mesotech, and finally just Kongsberg Mesotech.) The sale went through and Willy signed a five-year non-compete clause, promising not to work on sonar. But he did set up Imagenex a few months later when Alan Mulvenna joined the new venture. They both worked from home for the first couple of years. Willy honoured the non-compete clause, but that didn't mean he wasn't planning ahead. "We made interesting machines of other types, like a system to measure wood chips on a conveyor belt. And then from 1988 to '90, I taught myself how to make computers that were faster than any others at that time. When the non-compete clause expired in 1990, I started producing sonar using those computers."

His daughter Disa continued to be interested and involved in her father's work. In fact, there's still a note taped on the home office wall that says, "Dear Mr. Wilhelmsen, I am sorry to inform you that you have not paid one of your employees for the last week. You now owe this employee $7.50 as of today. Sincerely, D. Wilhelmsen, Secretary."

Many early Imagenex employees, like Gordon Kristensen and Jeff Patterson, had formerly worked at Mesotech and became partners in the

Helmut Lanziner recently joined key personnel at Imagenex. Left to right: Helmut Lanziner, Steve Curnew (sales manager), Gordon Kristensen (managing director and VP), Jeff Patterson (VP), Disa Wilhelmsen and Willy Wilhelmsen. Photo by Harry Bohm

new venture. Gordon remembers those early Mesotech days when things were invented on the fly: "Now, it's kind of like almost everything has been established, so it's not quite as exciting. Back then it was out of control!" Disa Wilhelmsen (marketing and graphics) agrees, "Nowadays, when we do have something new and exciting that hasn't been done in the industry before, it's much more controlled and confined. The sense of danger isn't there anymore."

Imagenex occupies a modest two-storey office in Port Coquitlam, every inch packed with equipment, assembly stations, test tanks, components and storage racks. While the company has a low local profile, that's not the case internationally. "Typically, our market is spread right around the globe," Willy states. "There's hardly any local business, although we've seen that starting to develop in the last few years."

What's obvious is that this company thrives on mutual respect and that everybody tackles their job effectively. Gordon adds: "Today it's so cliché to say 'thinking outside of the box,' but that's exactly what Willy and Helmut do. A big company is often so disciplined that its employees can't do anything beyond the normal conventions. All their people are going down the same path. When Willy does something, when I understand what he's doing,

I realize, 'Wow, that's not the way everybody else does it.' It takes his vision to see that there is another way of doing it. Often, it's a simpler way."

Like all companies in the subsea business, there are huge swings in the economic cycle. And Willy agrees that there are few bread-and-butter contracts to provide any kind of stability. "In all the years since I've started, there's never been orders just sitting waiting for us. We're basically working on a month-to-month basis. But eventually I just kind of stopped worrying about it. I try to have a business where we work regular hours, first of all. And I'm trying to keep the business small, so there aren't all kinds of office people. Once you start growing from 25 to 50 people, then the operation becomes totally different. So we try to hire what we need and not fire anybody. I'm very happy with all the years of work we've given people. Most of them have stayed with us a long time."

Mark Atherton, Can-Dive, Simrad Mesotech and Kongsberg Mesotech

Mark Atherton was a young buck when he got his scuba training, moved to the coast and took a job logging on Vancouver Island. Then he had what he describes as "an altercation with a tree" that nearly killed him. During the subsequent months in hospital, he had plenty of time to realize he didn't want to continue working in the bush. After watching several Jacques Cousteau documentaries, he decided to take a commercial diving course in Los Angeles. He blesses the doctor who cleared him to go diving, given the injuries he'd sustained. With his new credentials, he came back to BC, got into underwater monitoring with TV systems and eventually took a job with Phil Nuytten's Can-Dive. "I'm lucky I got into the sonar business when I did. There were no academic requirements back then and there was no training—you had to teach yourself. The only operational sonar book out there was Charles Mazel's *Side Scan Sonar Training Manual*, put out by Klein."

Mark is a decade or two younger than sonar pioneers Willy and Helmut, but he's equally passionate about sonar and has had the benefit of working with many of their inventions. During the years Mark worked for Can-Dive, it was often under the direction of Helmut Lanziner in the company's

Oceanographic Survey Division. On contracts in the Canadian Arctic, his primary job was operating underwater television systems, along with high-resolution seismic and echo-sounder systems. However interesting that equipment was, he was most fascinated by side-scan sonar.

"The two primary tasks for which side-scan is used are search or survey. Of those, search work is always a challenge, but when you find what you're looking for, it's the most rewarding. I'd get sent out on projects to look for planes, sunken vessels, cars and trucks that had been dumped into a waterway. You'd spot a target on the paper as it rolled off the

Mark Atherton, with his photographic gear, working for Can-Dive. Photo by Neil McDaniel, from personal collection of Mark Atherton

recorder and then try to do an interpretation. With only eight shades of colour and shape to deal with, it was difficult to get any detail. But sometimes that low-resolution information was enough. Considering the technology at that time, the side-scan records we collected were quite remarkable."

Mark gained more and more sonar experience working with Mesotech's various inventions. He recalls how in 1983 Mesotech came out with the first "compact imaging sonar," as it was called—the model 971. This high-resolution sonar soon became a standard for all ROV operations. In 1984 Mesotech garnered the "Special Meritorious Award for Engineering Innovation" for that invention, given by *Petroleum Engineering International* and *Pipeline & Gas Journal* magazines. "The moment I saw it in operation, I realized the sonar business was about to change."

Of course, the next challenge was to get it into the marketplace. Willy Wilhelmsen recalls the day when their rep, Terry Thompson, brought a potential customer from Texas to see this new sonar in operation. The system wasn't quite complete on the surface end, but Mesotech still arranged a demonstration at their test barge in North Vancouver. Despite the fact that there were still wires and circuit boards hanging loose, the sonar was deployed and produced impressive pictures of the bottom. The customer promptly said, "I'd like to buy two systems. What's the price?" Terry Thompson quickly quoted a price tag of $46,500 each.

"That was a surprise to us," Willy says, "because previous to the demonstration I had told Terry $36,500. But the customer was happy and only requested that we clean up the wiring a bit before delivery. Sure enough!"

The buzz about the new sonar took off in the oil and gas industry. Willy recalls getting a call from Phil Nuytten of Can-Dive soon after the first sales. Phil explained that there was a drilling operation offshore in Newfoundland using cameras to look for an abandoned wellhead at a depth of 1,800 ft (550 m). They had spent six days at a cost of $300,000 per day trying to locate it—with negative results. Phil needed to know if Can-Dive could find the wellhead with sonar.

"My answer was yes. Half an hour later, Phil called again and asked if we could be ready with an operator and sonar by six o'clock in the afternoon. Again, my answer was yes."

So Gordon Kristensen started looking for cables long enough for 2,000 ft (610 m). He located four 500-ft (150-m) lengths of conductor cable. Bob Asplin, Alan Mulvenna and Willy took a sonar system to the company's test site, along with the sections of cables and connectors Gordon had found. They trialled the whole system in the water; it worked just fine with the long cable.

Afterwards, Willy drove the equipment and Bob Asplin to the airport, where Phil Nuytten had arranged for a Learjet flight to Newfoundland (at a cost of $25,000). "The equipment was loaded onto the jet, and I started driving back home when I realized I hadn't seen the sonar head. I pulled in to the closest gas station and called Bob to see if he could spot the sonar head. It wasn't there." Willy then called Alan Mulvenna at home, and he went down to the test barge. There was the sonar head! He raced to the airport with it, and the plane took off late, but now all the equipment was on board.

The next morning at work there was a message on the Mesotech teletype machine saying that Bob had been successful. Upon his return, there were stories to tell. "Bob said that after finding the abandoned wellhead, he was a hero in Newfoundland. Even the taxi drivers knew his name." That trip resulted in the sale of the third sonar unit.

Mark Atherton recalls that Can-Dive also purchased one of these new side-scan sonar units for its ROV operations off the East Coast. He managed to lay hands on it when the unit was cycled through Vancouver. "Phil always told me I could play with any of the company equipment on the weekends as long as I kept it running. But there was no training manual for this new sonar. As a result, every free weekend, I'd head to Cates Park in North Vancouver, lower the sonar off the end of the dock and scan. No one was there to guide me through the data interpretation. I would just see a target, another target and another target on the CRT display—but what were they? To find out, I'd copy the positions of those sonar targets from the screen to my slate board, then get in my dive gear and try and find them in the water. I did that weekend after weekend."

That's how Mark learned how to translate the targets on the sonar screen to the actual objects in the water. He described it as a "fantastic learning experience." And once he had that information under his belt, Mark realized what an incredible tool scanning sonar was. Not only was it a fantastic instrument for ROV operations, but he saw that it was almost perfect for diving support, underwater construction and small-area search and survey.

"If I hadn't worked for Phil Nuytten, I wouldn't have had access to the sonars, submersibles and ROVs," Mark notes. "Sure, I kept things running for him and made him a lot of money, but in return, I got to play with everything. And I could *never* do that now. I couldn't walk into Kongsberg Mesotech [the current name of Mesotech's successor] today without that kind of previous experience and have the same opportunity. That's because as a company gets larger, they institute processes. And processes have to be followed. I get that. I even understand why. When you have to answer to the stock market, it's a completely different approach than if you're financing things yourself. Small companies can say, 'Yeah, we can do that.'"

Mark continued doing surveys for Can-Dive, but any time he'd run into a scanning sonar issue, he would head over to the old Mesotech building and talk to his mentor Bob Asplin. "He would always walk me through the math or provide the tech support I needed." In 1991, Mark received a call from Simrad Mesotech. Willy had sold the company to Simrad and moved on to other things, but Bob Asplin had stayed. Mark sensed a possible job offer—and he was right. Bob explained: "We have all these really bright people working here, but no one knows all the applications. You've used Mesotech equipment for all kinds of jobs, and we need someone with your experience. Are you interested?"

"Hell yeah!" Mark recalls. "I thought I'd do that for two or three years, but 27 years later my business card still has Mesotech on it, even though Simrad was bought out and the company is now Kongsberg Mesotech."

In the same year he joined Simrad Mesotech the company was developing a simultaneous dual-frequency side-scan, which was very advanced for its time. As the name suggests, two frequencies are transmitted at the same time. By looking at the two resulting sonar images, it's possible to observe subtle differences between them. With two different frequencies available, there might even be a target visible on one but not the other. "It was a game changer in assisting data interpretation," Mark recalls. "And the more I got involved in these ongoing sonar developments and worked with some of the best acoustic engineers in the business, the more exciting my job became."

Then came the development of high-resolution scanning sonar. Although it was eventually developed by Kongsberg Mesotech, the story starts back in 1983. Solving the problems associated with the original equipment reads like a technical detective story. It's also typical of how important on-site observations are in unravelling a problem—or a series of problems. "Back then," Mark explains, "scanning sonar had a transducer poking out one end of the head on a shaft. That shaft had a couple of o-rings to stop the water from leaking inside, and it worked great in shallow water. But as ROVs began operating deeper, the increasing water pressure caused leakage issues

Mark Atherton, shown here working on a 1993 side-scan survey of Vancouver Harbour for the Underwater Archaeological Society of British Columbia (UASBC). Personal collection of Mark Atherton

even though these sonars were supposed to have a 1,000-metre depth rating. After joining Simrad Mesotech, I found out that at the head's maximum depth/pressure rating there was a limited amount of time before the seal would fail. It wasn't a *might* fail, it was a *would* fail! This was happening with all sonar companies selling that product. And when a sonar head is worth $20,000 bucks, a customer tends to get justifiably upset when their investment turns into an expensive aquarium."

At the time, a UK company called Tritech had solved the flooding issue by encapsulating the sonar transducer in an oil-filled dome and separating the transducer motor from the electronics. Mesotech adopted a similar domed-head solution, but Mark noticed that images in Vancouver Harbour using the new domed system were not quite as sharp as the original design. The fuzziness became even more apparent when he took the domed sonar into the BC interior in December and later to a client in the Bahamas. But he didn't know why.

In 1998, Mark flew out to troubleshoot a particularly unhappy client in the tropics. Consulting with the company's engineering manager, both realized the problem had to be the oil. But, again, they didn't know why. Back home, temperature and pressure sound-velocity tests confirmed that sound velocity in oil changes dramatically with both temperature and pressure.

Solving that problem eventually required making the transducer longer, thereby allowing the beam to be more focused, and using a much wider bandwidth. It took some convincing, but in three weeks, the high-resolution scanning prototype was born. "The very first time I deployed the new Kongsberg Mesotech head at the test barge and looked at the images, I realized this was another game changer," Mark recalls. "Now our scanning sonar images were beginning to look more like photographs! A tire on bottom wasn't just a kinda', sorta' circular thing—you could see it was a tire. The detail was amazing!" But figuring out what the problem was and then how to solve it took a lot of years and expertise.

"To service your customer, you need the application knowledge, but you also need to understand what the client's data requirements are," Mark states. "I've always maintained that the sophistication of the electronics means zero. It's the data the system produces that must have value. People forget that we aren't selling a piece of equipment, we're selling data. That's the only reason that makes investing in the gear worth the expense. So I'm always looking at how to make the data better. It's as simple as that."

Passing on the knowledge

In 2011, Mark published *Echoes and Images: The Encyclopedia of Side-Scan and Scanning Sonar Operations*. It was a textbook he started back in 1997 and finished 14 years later. During those years, he did a lot of teaching, beginning with his first training course for the RCMP on surface-supplied diving (with air supplied by umbilical). That was followed by training Can-Dive personnel in underwater television systems and how to complete video surveys of docks and ship hulls. When Mark joined Simrad Mesotech in 1991 he also did field training, teaching people how to use sonar for diving support, construction, and for search and recovery (SAR) dives. While equipment always came with a manual, he helped bridge the gap between reading *how* to operate a piece of sonar and actually getting the data the customer needed. Mark began keeping notes with each class he taught.

About 1995 or '96, Mark recalls carrying a stack of training material and lecture notes to a presentation he was giving in Orlando when he ran into Bob Tether, a legend in the Police Dive Team world: "I had first met Bob in '81 when involved in the search/recovery of cars and people that drove off a washed-out bridge at M Creek just north of Lion's Bay. Using an ROV we located a victim and brought the body up to about 80 feet. Bob asked me to hold it at that depth and then requested that I look for a cloth bag or pillow case on board our vessel. He explained that if the head was likely to fall off, he'd need a bag to put it in so as to get the person's dental work. That was a first for me!"

At the conference, Bob spotted Mark's stack of notes and said, "You've got to get this into print. People in the field need this information." As Mark did more presentations, his small stack of papers became a big stack, and everybody asked, "When's the book going to be published?" Then one day Mark's father also asked, "Mark, what's the drop-dead date on this book you're writing? I'd like to see a copy before I go." Mark's father was 92 at the time, so his son got busy.

The big push for finishing the book took nearly 6,000 hours over the final three years, when Mark wasn't at his day job. Some sonars had never been written about before, but people in the industry proved very supportive in helping with data examples. "Clients and friends sent me records with the only requirement being to credit the images provided. It was all a verbal agreement."

The sonar textbook has been very well received. Mark especially remembers meeting two academics from South Korea on a boat. "One asked what I did in my spare time and my co-worker told them I'd written a book. It turns out that they were both PhDs, they both owned a copy of my book, *and* they both wanted their picture taken with me. That was a gratifying experience!"

Admittedly, a sonar textbook is not for everybody. In the book business, it fills a niche market, but Mark is well aware that the book has opened many doors. "Publishing it has allowed me to do training at a college level in the US, and the book is even being used at a survey school in China. When I arrived over there, I found 35 *photocopies* of my book. But the best news of all was that I was able to hand my father a copy of the book, dedicated to him, six months before he passed."

CHAPTER 10

Atlantis Submarines and underwater tourism

1983–present day

D ennis Hurd owns more submarines than many countries, including Canada. In fact his company, Atlantis Submarines, operates ten of them. But these subs are far from stealthy and certainly not for military or scientific use. Instead, they allow ordinary folks to gaze at the extraordinary wonders of the world's oceans. The stats are staggering: as of June 2019 his fleet had completed 560,467 dives, carrying over 17 million tourists underwater. Jacques Cousteau would surely have been impressed.

Dennis didn't start out dreaming of a submarine empire, though as a kid growing up in South Africa, he did enjoy the ocean. "My buddies and I went spear-fishing and diving and surfing, so it was kind of in my blood." But when he finished high school he was uncertain about what to do next, so followed his father's example, becoming a structural engineer. Eventually, he also did an MBA at UBC and got into banking. Both degrees would prove helpful when the time finally came to build submarines and finance a business.

After earning his MBA, Dennis went to work for the engineering department of the Industrial Development Bank (predecessor of today's Business Development Bank of Canada). His job was analyzing complex financing requests, particularly those that were highly technical. Although he doesn't recall the bank financing any of HYCO's work, during the mid to late 1960s he became fascinated with the company. "HYCO definitely had world-class technology. There was really nobody that could come close to them."

Eventually, Dennis Hurd switched jobs, joining HYCO in 1972 and managing the fabrication of the company's large new submersible, *Taurus*. HYCO was also where he first met Al Trice (engineering manager on *Taurus*) and John Witney. However, by the time *Taurus* was completed, the oil industry was shifting to ROVs. The intended market for HYCO's latest manned submersible evaporated.

Desperate to generate much-needed income, HYCO moved its submersibles *Aquarius* and *Pisces*, along with the *Hudson Handler*, to the Gulf of Mexico in hopes they could secure work surveying and inspecting pipelines. Dennis took over running that operation and oversaw a major contract

The fleet of Atlantis submarines provides an up-close and personal opportunity for all ages to view the wonders of the underwater world. Collection of Atlantis Submarines, Ltd.

working with the drillship *Discoverer Seven Seas* in the Mediterranean.

"But it all came to a grinding halt when HYCO had to shut its doors in 1979. At the end, Al Trice and I were the last two employees working. And after the receivers finished their games with us, we were left standing with nothing."

Cast adrift, Al, Dennis and Steve Johnson put together a little company called Southern Offshore Engineering. They got hold of *Pisces VI*, and a support boat, and one of Jim McFarlane's little ROVs. Then they headed back down to the Gulf of Mexico, using the submersible and ROV to monitor drilling operations for new oil wells. But by the late 1970s and early '80s, submersibles had become dinosaurs in the oil and gas industry. Once again, it was time to start over. Dennis decided this was his chance to pursue the idea of a tourist submarine business.

Building the idea

Initially, what really attracted Dennis Hurd to the idea of a tourist submarine was the fascination that people experienced going underwater, somewhat like the allure of scuba diving. "I remembered when we were running the HYCO submersibles and sometimes took a customer's representative on a dive. They got such a big kick out of going underwater, even if they might not get to see much of anything. I thought more about that when working in the Gulf of Mexico and even studied the idea of taking *Pisces VI* to the Bahamas for tours in the off season, but the numbers didn't work."

Now thinking of running his own company, Dennis concluded that he could make a go of it if he had a bigger sub that could carry 24 people. He dusted off his MBA and financial assessment skills and did market studies of the tourist industry. Simon Fraser University's marketing department also agreed to do a study for him. Armed with this information, it was time to start raising money.

But there were hurdles to clear. "When it's your own business, you still have to learn the hard way about how to raise money," he says. "And you can run aground pretty quickly when you're trying to find funding based on an unusual idea that has no parallels for comparison." Initially, Dennis invested his own savings. He came up with the hefty name of Sub Aquatic Development Corporation for his fledgling company, but wised up and shortened it to Atlantis Submarines.

The next stage was a design evaluation to see if the idea was even feasible. How much would such a sub cost? How long would it take to build? Where would he operate it? He did most of the technical feasibility himself, having done similar research for HYCO's *Taurus* project. "At this point I was still just considering a submarine conceptually, asking questions like, 'Is this going to cost ten million dollars or two million?' I also had to digest all the market studies and come up with a plausible development plan."

Then it was time to talk to his family—his three brothers, his sister and his parents—about financing the next stage, the technical side. The actual design of the sub generated numerous details that had to be figured out. That meant getting a lot of consulting people involved. Soon there were pressure vessel consultants, and fibreglass consultants, and structural consultants. Guys like John Witney, Jim Warren and Tom Roberts came on board. Slowly Dennis built up his design team.

Each new stage of the idea required more money. Sourcing funding became a recurrent exercise. That meant getting friends of the family interested, and then the friends of friends. A critical factor was that the Government

Atlantis I under construction in 1984–85 at Vancouver Mill, on Vancouver's False Creek. Collection of Atlantis Submarines, Ltd.

of Canada had a business investment program at the time, so if a research and development project qualified, the investment generated a tax credit. That made a big difference and certainly reduced the risk for new ventures like Atlantis.

Finally, in 1984, it was time to actually build a submarine. Although Dennis was elated, there were still important details to figure out. In particular, where was the sub going to operate? Many locations were immediately ruled out because they were too seasonal. "A high-cost business like

An important evaluative step in planning and fabricating the initial *Atlantis I* was a plywood mock-up of the sub's interior, one that would facilitate passenger access and viewing. Collection of Atlantis Submarines, Ltd.

Atlantis I arrived in Grand Cayman in November 1985. Collection of Atlantis Submarines, Ltd.

this succeeds only if there's a flow of tourists pretty much year-round. Sure, you can tolerate maybe one month a year where you're really losing money, but that's about it. It's not just money coming in to consider, but money going out. So you need lots of tourists all year because your fixed costs are so high."

Besides operating costs, there are crew considerations. It takes a year to train and certify a pilot, and it can take a senior pilot several years to know the ropes. An operations manager requires maybe five years of experience. "So you can't just lay people like that off. That's been a problem with lots of tourist submarines—they can't stand the seasonality. Ten subs tried to make it in the Mediterranean; they all failed because the market is very, very seasonal."

Besides the challenge of finding a year-round site, Dennis ticks off other critical considerations: "First off, you need about a million tourists nearby, but you also need fish and coral so passengers have interesting things to see. And you need good visibility, otherwise dives will be cancelled. You need protection against storms. As well, the current can't be too strong or the surface too rough because we transfer passengers from a boat to the submarine. And some things you learn as you go. For instance, we learned that bussing tourists too far to a site doesn't work."

But the list doesn't stop there. The political environment needs to be stable and amenable to business because various government permits are essential to operate where a firm hopes to work. Most islands don't want a foreigner coming in and running a business without local involvement, so the population needs to be big enough to produce the pilots needed, as well as training programs for them. Until locals are trained, expensive work permits must be obtained for outsiders; that includes initial pilots, managers and accountants. Working through all these considerations, Grand Cayman Island in the Caribbean rose to the top of the list for the first Atlantis location.

While Dennis was building the submarine back home, he also began working with the Cayman government, a very long process. He learned the necessity of keeping at least one if not two people on site all the time. "If you didn't keep a presence on the island, your application ground to a halt. Eventually, we did get all the permits required, including a duty-free entry, which is a big deal. We started over 30 years ago in the Caymans, and now all our pilots are local wherever they are. Occasionally we might have one work permit out of 20 or 25 people, but that's it."

With the success of the first Atlantis submarine in the Caymans, the company began thinking about where else it might open. Barbados seemed likely. "That plan was spurred on by the fact that Expo 86 was coming to Vancouver," Dennis recalls, "and the organizers said they would like to show off one of our subs. We were building *Atlantis II* and accelerated its construction because the chance for that kind of exposure was rare."

St. Thomas was another Caribbean island on the list of potential locations. "At that time, St. Thomas had about five million tourists arriving on cruise ships, a massive number. Based on our results from the Cayman Islands, which indicated that about 20 per cent of all tourist arrivals took a ride on our subs, we went to Toronto and with Merrill Lynch raised about three to four million to build the next sub and develop St. Thomas. Gradually, the Caymans settled down to a more realistic 10 per cent, depending on the hotels and cruise ships, but the initial figures suggested that St. Thomas would do really, really well." But that was on paper. In reality, it was a different market, because the cruise ship folks emphasized duty-free shopping to their passengers, not going for a ride on a submarine. "The hard lesson to learn about St. Thomas was not to get too cocky. I wish we'd been smarter there because we lost a lot of money. So you *always* need to do your background research. You can't make assumptions."

Some locations worked, some didn't. Atlantis tried Cancun but faced a demand for extremely high commissions, so the firm went to Cozumel. Also, not every base was the same cookie-cutter operation. "In the case of Guam, we actually sold them the submarine," Dennis explains. "It's under licence to them, so we get licensing fees and in return we provide all the technical support. And they've done really well."

Atlantis Submarines currently has subs in Grand Cayman, Barbados, Aruba, Cozumel, Guam and the US. Certainly Hawaii is the company's biggest market, with five of the Atlantis fleet stationed there—that's one each in Kona and Maui and three in Waikiki, including *Atlantis XIV*, the world's largest passenger submarine, accommodating 64 passengers on each dive. Since there are very few cranes big enough to lift it, the sub is based in Honolulu, where there are dry docks that can handle it. The rest of the Atlantis subs carry 48 passengers.

Atlantis XIV, the world's largest passenger submarine, was built in Everett, Washington, and launched in 1994. It operates in Waikiki and can accommodate 64 passengers. Collection of Atlantis Submarines, Ltd.

"People often ask why Atlantis doesn't operate in Vancouver, since the city gets about the same number of tourists that Hawaii gets," Dennis notes. "But Vancouver visitors only come from May to September, and we can only carry so many people a day. There are beautiful dive sites on Vancouver Island, for sure, but what do you do with your employees and your submarine the other six months of the year, when there are no tourists?"

As word of Atlantis's success began to spread, other companies started to sniff around, thinking tourist subs were a good deal. "However, they only thought about building subs and then selling them to people who had little experience in operating them," operations manager Les Ashdown notes. "There's just not a lot of money in building a tourist submarine, and there's only a limited market for them. Over the years, Atlantis has done it better than anybody else because we were the only ones that built them, owned them *and* operated them."

Other companies even asked if Atlantis could run their fleet and just pay them a percentage. The answer was no. "Their subs were very complex, as well as difficult to maintain," Les adds. "Ours are literally a McDonald's hamburger style of operation. Our engineers settled on a simple design, and that was lucky because in the beginning we didn't know how important it would be to turn out a very simple, but safe, vessel that was easy to maintain. The difference for us was that we operated our subs, so we learned from that, too. Today we own most of the world's operational tourist fleet."

However, all of that careful planning and practice couldn't predict the staggering blow that Covid-19 would deliver to worldwide tourism-related businesses, including Atlantis's global fleet. After a total shutdown, the company is happy to be re-opening operations, beginning in Hawaii, as of 2021.

Pilot certification

Building subs is one thing and training pilots to operate them is another. "No matter what the subs are used for, pilot training should be rigorous," Les states. "Ours certainly is." Learning to pilot, along with learning maintenance and other subjects, requires a program of study in a variety of areas, coupled with experience. Not surprisingly, Atlantis wrote or updated a lot of the training manuals in those subject areas.

Les details the logistics: "In terms of timing, if Atlantis hires you as a co-pilot, it will take you about 14 months to become certified as a pilot. You go to a site and work through the co-pilot program, module by module, signing off on every task after you have accomplished it. Then when you have finished all 12 manuals, including seamanship, safety, maintenance and others, you write all your tests, including all the exams. If you pass those, you get a certificate from us. With that, the US Coast Guard presumes you're a certified pilot. They have an exam you take, too, but it's only 100 questions and it's a computer-generated exam taken from 1,000 questions out of our manuals. When you pass their test, you're issued a certificate from the US Coast Guard stating you're a licensed submersible pilot, but it's only for a specific location. If you want to go from piloting in the Caymans to Barbados, for example, you have to do a dozen dives and re-certify again for that new location. It's a tough process, but those who go through it certainly know their stuff. Also, in order to pilot our fleet of subs operating in the United States, you must be a US citizen. For the other sub locations, that's not a requirement."

Dennis talks frankly about the business. "These days we're not building more subs for the Atlantis fleet because we haven't identified a good new site." The tropics are idyllic, but there are high currents to consider and the weather can be a wild card. Atlantis has hurricane plans for all the sites, which include storm moorings specially designed for their subs and ferry boats. Even lesser weather problems can interrupt business. Dennis adds: "If you get too much bad weather, then the agents say, 'I don't want to sell your product because I keep having to give refunds to people.' So weather is definitely a factor."

And then there's insurance. Initially Atlantis had trouble finding an insurance company, but the tourist submarine world is accepted now. "Of course, if we'd had an accident back in the early '80s with the first submarine, the company probably would have died right there," Les admits. "Fortunately, in 30-some years of operations we've never had any serious accidents with passengers, and that's carrying over 17 million people on 560,467 dives!"

Building the Team

John Witney

John is engineering general manager for Atlantis. Prior to that, he was project engineer with HYCO. "One huge advantage of working with a small company like HYCO was that I got to design submersibles and then be alongside the people building them. Not only that, I also got to test the vehicles and then actually go and work them in the ocean. A lot of us at HYCO were also trained pilots, so we'd dive in the North Sea and off the coast of Africa, the Gulf of Mexico and the east coast of Canada. We set a bunch of records doing that. In fact, I was the first Canadian submersible pilot, with Tom Roberts as my co-pilot, to ever go down to 6,500 feet in a submersible—and get back to the surface!"

After HYCO shut down, John set up his own consulting company for a couple of years. Then Dennis Hurd phoned, asking him

John Witney (left) and Les Ashdown (centre) brought their extensive maritime background and expertise to the company, joining company president Dennis Hurd (right) for decades until their recent retirement. Collection of Atlantis Submarines, Ltd.

to come on board as a consultant. "I liked what he was doing. It was something I believed in. I had worked very hard helping design the *Pisces* subs and going deep with them, and I wanted the public to see the underwater world, too. Also, we're spending billions of dollars going up to that speck in the sky, but we've hardly taken one step forward to explore the ocean which is right at our feet. I wanted to expose people to that.

"The idea of a tourist submarine wasn't new. But when Dennis wanted to do it, I thought it was a great opportunity—albeit one with a whole bunch of challenges. When we started, we intended to take 28 people down in a submarine, but first we had to build it *and* get it certified. At that time, the American Bureau of Shipping and the US Coast Guard were not at all familiar with tourist subs." So for years John has worked with the ABS's special committee on Underwater Systems and Vehicles, helping develop the rules that should apply to subs like those Atlantis operates. He also works on the PVHO (Pressure Vessels for Human Occupancy) Standards committee. "Every organization needs rules to guarantee safety, but the biggest thing is making sure the rules make sense and aren't just rules for the sake of making rules. That's just one of the special challenges I enjoy about working for Atlantis!"

Tom Roberts

Tom also started his subsea career at HYCO, gaining vital skills and practical experience in the shop and in the field. "This whole subsea thing is all about the learning. You never stop. That's what's great about it, that's the challenge." After HYCO folded, Tom briefly worked for Southern Offshore Engineering, which Dennis, Al Trice and Steve Johnson operated in the Gulf of Mexico. Then Dennis came up with the idea and the funding for tourist submarines. When Atlantis was ready to start building them, he called Tom. "I became manager of electrical engineering, and then I was manager of assembly for all of the Atlantis boats. That was 20 years of my life.

"But eventually there were no more viable spots in the world for Atlantis to put a submarine, so the company quit building subs. At first, they were just going to do underwater tours, but then the company got into buying other businesses in the tourist industry and making them work. Atlantis pretty much created the tropical tourism market. I left only when there were no more submarines to build."

Les Ashdown

Les started at HYCO like almost everyone else. He remembers taking guys from the Fisheries Department or the military down in the *Pisces* and how they were absolutely fascinated by the experience. "We used to say, 'You know, we should get people to pay for a ride in one of these things.' We just talked about that, but Dennis actually did it."

During a slow period in the early 1970s, Les moved from HYCO and joined Lockheed Petroleum Services as their first "capsule operator." Essentially, he piloted a very large, atmospheric, non-pressurized bell that accommodated a crew of four and mated to a bottom chamber on the live oil well. The life-support equipment and training required to put people on a live well at depths of 2,000 ft (610 m) below the surface of the ocean were extensive.

Then Les came back to Vancouver and managed Lockheed's fabrication plant on Annacis Island, in the Fraser River. As operations manager, work regularly took him to Brazil, the Gulf of Mexico and the North Sea, as well as reporting to Houston one week out of every month. He recalls, "No matter what you did at Lockheed, you were just one part of a huge company. I was employee #29206."

Then the subsea world moved on to robotics and ROVs. "When Lockheed finished up its last oil and gas contract in Brazil," Les recalls, "I had to lay off everyone—114 people. I was the last. My boss in Calgary called and said, 'What are you doing for lunch tomorrow?' I said, 'Probably meeting with you.' He flew out the next day and brought along a very attractive severance package, so I took it. As we were finishing up lunch, I had a startling realization. 'The company just gave me a brand-new company car, but now I don't have a way home.' 'Not a problem,' he said. 'It's yours.' He gave me the car! Those were the days."

At the time, Atlantis was looking for an operations manager. Tom Roberts said, "I know one," and called Les. "I came over for an interview and started the next day. That was that! All of it—at HYCO, at Lockheed, and at Atlantis—was learning on the job. I've been lucky for my whole working life because I've rarely had a day when I didn't want to get up and come into work. You can't beat that!"

Mavis Mitchell

Mavis is like many subsea employees, moving from project to project and company to company as the opportunities arise. However, being a woman in the subsea industry was, and still is, far from typical. She first earned a Red Seal journeyman's certificate in the trades and then went to BCIT for engineering training, particularly in mechanical design.

"Back then, you looked in the newspaper for jobs. I spotted an ad that said, 'Are you a mechanical or electrical technologist? Do you like water sports?' Well, yes, I liked water sports, having grown up on Christina Lake. 'Would you like to live in the Caribbean?' I thought, 'Gee, that sounds just like me!' So I applied for the job. What Atlantis Submarines didn't put in that ad was, 'Do you want to build submarines and pilot them?' Out of the 400 applicants, I was the only woman that applied. They short-listed me for an interview and sent me a letter saying the job was actually to be a submarine pilot and maintainer. I remember thinking, 'What does that have to do with water sports and living in the Caribbean?'

Mavis Mitchell has worked for many subsea companies, beginning with Atlantis. Currently she is an instructor at BCIT.
Vickie Jensen photo

"It took me the entire weekend to convince myself that I could do that kind of work. So I went in for the interview, but they were quite leery about hiring a woman because most of the men they were hiring were rough and gruff commercial divers. They didn't think I could manage that. Heck, I'd already been through *that* fire."

In the end, Mavis didn't get the job. But Dennis's executive secretary decided that Mavis *should* become a pilot. It was time. "She really started my career," Mavis recalls. "Whenever Dennis was in town, she'd phone me and say, 'He's in town. Maybe you want to get in touch with him.' And I did that. So when Atlantis went to build their second submarine, they called me back to talk."

Mavis went in for another interview and the company took her on for a trial build, working on their second tourist submarine under construction in False Creek. That went so well that Atlantis hired her. Mavis worked in the office, doing the design work that she especially enjoyed.

When the sub was done and went off to Barbados, Mavis went along. She recalls, "Barbados is beautiful, but it's small. The population was only 250,000, which was even smaller than Vancouver back then. I still wanted to get a bit more education, so after two years I came back to Vancouver and returned to BCIT for some electronics courses, and also to Vancouver Vocational to get more proficiency in the electronics side of things. After that, I came back to Atlantis and did design work on another of their submarines."

Between subs, Mavis took some time off, returning when the company began working on a 48-passenger sub for Hawaii. Because of the Jones Act in the US, the US-based subs had to be fabricated in the States, so Mavis went down to Everett, Washington, and worked there. "Through all of this, I got to work with Dennis and Les and John and Tom Roberts. They were pivotal to what was going on back then. And the work was so interesting. I think I turned out good for Atlantis, and I think that they enjoyed hiring me, so it was mutual."

Ben Hurd

Ben is Dennis Hurd's oldest son. At 42, he's the CFO of Atlantis. "I was around 10 years old when Atlantis was getting started. Dad travelled quite a bit, especially every time they started a new site. But whenever he was back in town, he made time for family.

"I went to UVic in business. Around 2002, I started a venture providing organic groceries to Dairyland's home-delivery program. After university, I thought about going into the family business, but I wanted to make my own way, do my own thing. My thinking was that if I did come into Atlantis, I wanted to be able to bring something *to* the company.

"So next I joined PriceWaterhouseCoopers and worked in Vancouver with them until 2009, completing my Chartered Accountant designation. The work involved lots of travel, but I learned a lot. Then I moved to Victoria, working for the Auditor General of BC from 2009 to 2013. I was involved on the financial side but also doing performance audits, all of which helped to build up accounting experience. Around that time an opportunity came up at Atlantis, and I was excited to join. The business has a long history and lots of excellent people. Working with family has been great. We're all pretty easy-going so we get along well.

Ben Hurd and Dennis on an Atlantis sub in Cozumel, 2020. Collection of Atlantis Submarines Ltd.

"Financially, this is a business with ups and downs. It helps to have geographically diverse operations, so hopefully if one or two sites are struggling, the other sites are doing OK. But it can be tough. Everything could be going fine, then a hurricane comes along and causes all kinds of problems. Or in the case of Hawaii, when the Kilauea volcano erupted in 2018 on the Big Island, it hurt our operations in Kona.

"Our last submarine was built in 1997, and currently there aren't any new sites where we think a submarine might be viable. We're hoping that will change, so we regularly talk to the general managers of the sites, and they update us about new ideas and possibilities as well as the day-to-day operations. The tourism business does change. For example, there are islands in the Caribbean that may develop their cruise-ship market or there may be opportunities for expansion in areas like Asia and South America that weren't feasible 30 years ago. In the meantime, the focus is on maintaining the vessels and submarines and delivering the best experience as possible at the current sites."

Daniel Hurd

Daniel is Dennis Hurd's younger son. At 32, he's an engineering specialist in the engineering department. "Currently I work with John, Les and Alastair, providing technical support and ensuring that everybody is on the same page. I also visit our operating sites yearly, surveying the subs and working with the teams there. Of course, these are nice places to visit—no complaints about that—but we spend a lot of time working at the dock.

"I received my civil engineering degree at BCIT. Before that I spent three years working and training to be a pilot with Atlantis in the Cayman Islands. I had no intention of actually returning to the submersible business when I received my civil engineering degree; I was more focused on the construction industry. But Atlantis thought it would be a good idea to bring my operations experience and engineering knowledge to the table. It's invaluable to have both the theoretical knowledge that comes with a degree and practical experience. That's why I still maintain my pilot's licence. It helps with credibility in the industry and maintains a connection with the offshore crews.

"I am proud of the history of the submersibles industry in my hometown of Vancouver and the impact Atlantis has had. I remember watching these subs being built as a kid. My dad would take me to the warehouse, where there would just be a big steel hull. Vancouver was, and still is, a large hub with a lot of subsea tech history.

"Worldwide, we have around 400 employees, with 11 in the Vancouver office. A lot of people working here are 20- to 30-year employees. Both our culture and our work ethic are long term, and we're more like a big family. Supporting our existing operations keeps us busy, and we've got a pretty rigorous preventative maintenance program for our subs, too. As well, the submarines themselves are continually evolving because the equipment on them keeps changing or updating. We source a lot of these parts locally and ship them out worldwide."

As pilot, Daniel was proud to take his parents on a submarine tour in the Caymans. Personal collection of Daniel Hurd

Working at Atlantis is not your typical job. And that's true for those in the office, as well. Les Ashdown's official title at Atlantis is operations manager, which means he looks after field operations as well as engineering and purchasing. John Witney is engineering manager, Arnold Point is purchasing manager, and Alastair McGilp is senior designer. Aside from their regular duties, all four at Atlantis are on call 24/7. That means even if they're elsewhere, they answer all emails and phone calls coming in from the sites every day.

Les explains the reasoning for that kind of monitoring: "The big factor is that there are many things in the ops manual that will not allow a sub with any sort of problem to go diving that day without engineering approval. It can be a thruster that's out or a valve that isn't functioning correctly. And even though there's a technical problem, there are still passengers standing on the dock waiting, and Atlantis doesn't want to cancel that dive if they don't have to. So the company has a regulation with the insurance company that says if the site can get ahold of someone in engineering and review the problem, and if they say it's still safe to do the dives for the day or for the next couple of days or even just that dive, then they're free to go out and still be covered by insurance. But without that engineering OK, there is no insurance. So year-round monitoring is crucial because the subs dive year-round."

John Witney adds: "When us old-timers finally retire, things will change a bit because the younger ones don't have the background that we have. For example, I just came back from doing a survey in Guam. When the US Coast Guard surveyors asked questions, I had the background to be able to explain, 'This is the way it is', and give them the reason why. But the next generation is not going to have that same practical expertise. Fortunately, Dennis's son Daniel is working with us now and he's also on the ABS board with me."

"When we hired people to build our submarines, we looked for people who had the motivation, the interest," John continues. "That's who we ended up hiring, not necessarily those who had degrees. We trained them and they became really good. We'd also bring in vendors to explain their products. As a result of that training and on-the-job experience, our employees also ended up being in high demand around Vancouver."

CHAPTER 11

John Horton's elusive dream:
The *Auguste Piccard*

1970–1983

N ot every entrepreneur in BC's subsea industry started out working in the water. John Horton was much more of a dry-land investor and enabler, thanks to family money. He was used to paying to get things done rather than doing them himself, so never quite understood the nuts and bolts of designing or maintaining a subsea vehicle. Horton had "almost" completed a PhD in physics and was inclined to dispense technical advice (often unwelcome). However, his funding *was* crucial to the start or survival of several BC subsea companies. Unfortunately, his dream investment nearly ruined him.

Upon his father's death, the family fortunes did not pass to John Horton, because family policy evidently dictated that inheritance went to "every second generation." However, as part owner and head of Chicago Bridge and Iron, he had access to corporate finances of the family's giant construction corporation. It was commonly noted, with some irony, that "CB & I didn't build bridges, didn't work with iron, and wasn't based in Chicago." The company was centred in Maryland, with maritime access for building offshore oil rigs, among other things. During the 1960s, John Horton came to BC, where he was listed as a "Vancouver-based businessman." His wife and children lived at a distance, in Victoria.

In 1970, after briefly considering a tourist submarine business with HYCO, John Horton bought two historic submarines—the *Auguste Piccard* and the *Ben Franklin*—and established Horton Maritime Exploration in Vancouver. He also would make various investments in Canadian underwater tech companies, buying into Kinetic Sciences Inc. (KSI), RSI Robotics, Ballard Research

John Horton came from a moneyed family who had built the Chicago Bridge and Iron empire in the US. But he had little understanding of marine construction himself. This light-hearted photo was taken underwater on the *Auguste Piccard*. He was hoping to get into the Guinness Book of Records for the deepest Frisbee game.
Personal collection of Guy Immega

and the sonar company Mesotech Systems, among others. Some of these ventures, such as his funding of Mesotech, were quite profitable (see Chapter 9). Even more notable in terms of a successful investment profile was Ballard Research (now the globally known Ballard Power) which got its start under Horton's funding umbrella when Geoff Ballard came to Vancouver to do research and development on high-energy lithium batteries. However, not all of Horton's investments were successful— some broke even and a few were dogs. The most calamitous was his own Captain Nemo dream that drove John to purchase two submarines. Ultimately, this dream would prove to be a personal financial nightmare.

Guy Immega, engineer and owner of Kinetic Sciences in Vancouver, first met John Horton in 1991 when he convinced Horton to invest in KSI prior to the award of an initial $1-million contract with the Canadian Space Agency (for autonomous control of Space Station robots). Immega eventually paid back that investment and then later went on to write about Horton and the *Auguste Piccard*. But the real payoff was the opportunity to know the man: "John Horton had a booming voice and stood about 6 ft 5 in tall, so he had a huge physical presence with an ego to match. He was somewhat of an inspiring figure, if you wanted to believe what he was doing." Reputed to be a billionaire, anyone who knew John Horton also confirmed that he was a notorious tightwad.

Besides being the titular head of a major corporation, John Horton also dreamed of owning his own profit-generating submarine. More importantly, he had the funding to make it happen. So in 1970 Horton bought the famous tourist submarine *Auguste Piccard*. In 1964 it had taken some 32,000 visitors at the Swiss National Exhibition down to the polluted depths of Lake Geneva. Later, Horton purchased the *Ben Franklin*, another historic submarine that had made the first and longest drift dive in the Gulf Stream current in 1969.

The *Auguste Piccard* was named for the famous Swiss physicist, renowned for having achieved both record ascents into the stratosphere as well as record descents into the ocean. After the Swiss National Exhibition finished, the submarine was advertised for sale for a million dollars. John Horton acquired it for only $100,000—a steal of a deal! Now he owned the largest non-military submarine in the world, one also capable of deep diving.

The next challenge was to refit the sub after years of lying idle and then find work for it, ideally in the lucrative subsea oil and gas industry. Originally, Horton intended to do that at Chicago Bridge and Iron's shipyard on the East Coast. But the US Navy had a nearby nuclear submarine base and refused permission to launch his submarine in those waters.

Instead, he loaded the sub on a freighter and brought it to Vancouver. His intention was to take it to Star Shipyards, since he had an interest in that company; but the shipyard inconveniently went bankrupt. So Paul Howard, general manager of Horton Maritime Exploration, facilitated getting the *Piccard* into Dillingham Shipyards in North Vancouver. Horton told the board of Chicago Bridge and Iron that the submarine's long and costly refit would provide a means of inspecting the offshore drilling platforms the parent company had built. He must have been persuasive, because the Canadian government also invested a million dollars in the project.

The *Auguste Piccard* required an extensive refit, including the addition of crew quarters, diesel power and equipment for sonar and navigation. Horton hired local BC workers. One of them was Tom Roberts. "Horton decided that if the sub was properly instrumented, it would make an excellent platform for doing routine bottom profiles for oil and gas companies who were exploring new well sites. And it was a good idea since places like the North Sea have huge surface waves." Extreme wave action often introduced weirdness into the readout of towed instruments such as magnetometers, sub-bottom profilers and side-scan sonars. Horton's whole idea was to get rid of this surface wave influence by mounting whatever equipment was needed on the sub, which could travel a straight line a thousand feet below the turbulence. The result, he was sure, would be much better, cleaner data.

"Furthermore," Tom Roberts continues, "he would be able to sell this idea to every oil company on the face of the earth, everybody would be happy ever after, and his company would make a lot of money. Well, it didn't quite work out that way. Nobody in the gas and oil drilling industry was interested in it." Also, in the '70s, other non-manned systems for locating oil wells were developing, so a submarine close to the ocean floor wasn't an advantage any more.

Initially, Horton held fast to his dream as the refit got underway. In 1975, Danny Epp joined the project to handle engineering on the sub's entire electrical system. He recalls that the *Ben Franklin* was in the North Vancouver shipyard as well and had already had been stripped of its more state-of-the-art sonar equipment for installation on the *Piccard*. The sonar company Mesotech was in the same building so did general acoustic work for the *Auguste Piccard*. Mesotech founder Willy Wilhelmsen recalls: "We were maybe the only company supplying equipment that made money on the deal. Even so, I didn't like John Horton from day one. He used to show up and try to give us technical advice. He thought he knew more than anybody else, but his ideas were totally screwed."

Danny Epp also had to contend with John Horton's assumptions of expertise. "He had a bee in his bonnet about a particular type of wire, called pyrotenax, that's normally used in explosive atmospheres. It wasn't at all helpful for our work because the wire was stiff, difficult to bend and not at all accommodating to a submarine hull. It was a mechanical nightmare, but Horton insisted. He was a bright guy but didn't know enough to make the right decisions and had no experience at all with the marine environment."

Horton had also met Geoff Ballard and decided that a lithium battery was the key to longer endurance for the sub. When that didn't materialize, the refit went ahead with the lead-acid batteries that were typical for all conventional subs at that time. In order to become a self-supported ocean-going vessel, the *Piccard* "got loaded with as many batteries as it could float, along with two diesel generators in order to charge them. But in the end, the sub wasn't self-supporting and required a surface vehicle for supplies, towing to site, and support crew," Danny Epp adds.

The sub's extensive refit took two years and sucked up money. In the end, Chicago Bridge and Iron's board booted John Horton out. They gave him the sub as a parting gift, but that was all. Suddenly, he was without access to the company money that had provided his cash flow. Finding work for the *Auguste Piccard* became critical.

———

A two-day job with BC Hydro, scouting a submarine cable route, was the extent of any local work that Horton could uncover. And he was fast running out of money. At the hint of a contract with the US Navy in San Diego, Horton directed the sub and its crew to head to California, captained by Tod Slaughter. He was a renowned submarine commander who, unfortunately, had taken to drinking. Some crew members, such as mechanical engineering technologist Brian Gorbell, stayed with the sub as unsung heroes; Brian eventually became operations manager.

If the US Navy didn't pan out, Horton's back-up plan was to transit through the Panama Canal to the Caribbean, where Horton Maritime would have access to the Gulf Coast offshore oil rigs. Surely there he would make his fortune.

However, Horton couldn't afford to put the submarine on a freighter. So the crew would have to sail the *Piccard* from Vancouver all the way down the Pacific coast to San Diego. In order to make the voyage as economical as possible, Horton outfitted his submarine with a mast and sail. The mast was bolted to the deck and could be folded down onto the conning tower when the sub was underwater. Topside, the mast could be winched up and a giant Genoa sail hoisted to make way sailing, or at least to assist with motor sailing. When on the surface, the sub's two new diesels turned a generator that charged the bank of batteries. In turn, they powered the electric motor that drove the propeller. Despite these adaptations, the *Auguste Piccard* only made about four knots in the open ocean. When the sub finally reached San Diego in 1978, the

Having sailed from Vancouver, the *Auguste Piccard* fell into disrepair over a period of several years in Mobile, Alabama. After Horton signed a contract with Sea Search Armada, the sub was dry-docked there in hopes of being brought back to working order. That didn't exactly work out as planned. Personal collection of Danny Epp

Navy came aboard for a quick look and then left. They weren't interested in a converted submarine that was essentially a very costly curiosity.

Following his backup plan, the sub and its Canadian crew continued south, with John Horton directing operations from Vancouver. Brian Gorbell's ingenuity managed to keep the diesels and hydraulic steering operative despite the corrosive salt air and tropical heat. Eventually, the crew got through the Panama Canal and reached Mobile, Alabama. There they ran out of everything, including money and hope. To save moorage costs, Gorbell found a bayou and tied the sub to a tree in a swamp estuary. It lay there for two years, mouldering in the sweltering Alabama heat as its captain allowed the sub to fall into disrepair.

Just when the *Auguste Piccard* seemed destined to become a derelict, Horton managed to snag a contract with Sea Search Armada in 1981. The

The support vessel *State Wave* was hired to tow the *Auguste Piccard* from Alabama (that departure was subsequently revised to Mississippi) to Cartagena, Colombia. The *State Wave* remained in Cartagena with the submarine for several months before returning to the US without the sub. Personal collection of John Swann

investment group was focused on finding lucrative historic shipwrecks—some would say treasure hunting. Principals Jim Banigan and Warren Stearns had both recovered sunken treasures before and presented a plan to mount a search for the legendary *San José*. It was reputedly the most valuable sunken treasure ship from the Spanish Main. Sunk in 1708 by a British naval ship off the coast of Colombia, the ship's rich cargo had never been recovered. If the *August Piccard* could locate that treasure, Sea Search Armada, Horton and his crew would be rich. The catch was that none of the investors would get paid *until* they found the treasure.

But first, the sub needed to be brought back to working order. Again, Horton wisely recruited good help from Vancouver, beginning with Helmut Lanziner as project manager. But the job wasn't exactly as advertised. In fact, when Helmut arrived in Mobile he realized that the *Auguste Piccard* hadn't been in water any deeper than 10 ft (3 m) for several years. Consequently, the first thing he did was an assessment of what needed to be fixed before the sub could go to sea. He came up with 212 items. John Horton was incredulous. "I've had a crew down there for several years. What do you mean?"

Drunk or sober, the sub's captain hated Helmut, questioning both his expertise and authority. Nonetheless, a refit was necessary if the sub hoped to get to Colombia. There was no shortage of problems. The propeller shaft leaked on dives, squirting seawater into the engine room and the sub's electric motor. The dive ballast tanks worked sometimes—and sometimes not. On early dives, smoke filled the sub. Captain Slaughter's response was simply, "well, the seals need a bit of work, I guess." There were always plenty of excuses. However, the crew managed to get 30 or 40 of the 200-plus systems working.

In order to get to the job site as expediently as possible, a rental support vessel called the *State Wave* was hired to tow the submarine to Cartagena, Colombia. Towing would be almost twice as fast as the sub could manage on her own power. But first the *Auguste Piccard* needed to be certified by both Transport Canada and the ABS regulatory bodies. This required a test dive to a depth of 500 m (1,640 ft) with both inspectors on board. In order to save time, the dive was scheduled at an appropriate water depth on the way to Cartagena.

Helmut had returned to Vancouver to deal with his own company business, so missed the certification dive. When coordinating rendezvous plans for his return, he called on the radio to ask how the trials had gone, but the *State Wave* crew asked him to wait to discuss things. Knowing that any conversation on the radio telephone could be heard by other vessels in the area, Helmut immediately feared there had been "issues." And he was right.

With the Canadian and US inspectors aboard, the *Auguste Piccard* had begun the scheduled descent to 500 m, but the main vent valves, badly neglected during the sub's years of idleness, could not cope with the high-pressure air necessary to blow the main ballast tanks. As a result the sub descended rapidly, stern first, much deeper than intended. The captain managed to get the sub back under control and to the surface, but only after dropping the 6.34-ton emergency dropweight. Since the sub couldn't sail on without one, the immediate problem was to find a shipyard capable of making and installing another drop weight. Back tracking, Horton got the sub into the Chicago Bridge and Iron shipyard in Pascagoula, Mississippi. The sub got another dropweight along with further delay and major expense.

During the US shipyard stay, Helmut was advised that once the sub reached Cartagena, it was customary to "gift" a case of Scotch and a case of wine to the appropriate officials. Knowing Captain Slaughter's history of drinking, he waited until the night before departure, then went ashore, bought the requisite liquor and wine and hid it under his bunk. Finally, the *State Wave*, with *Auguste Piccard* in tow, left Pascagoula late on October 16, 1981, heading for Cartagena. However, two bottles of Scotch quickly went missing. Captain Slaughter was drunk, and Helmut was fed up. He sent a message to Warren Stearns, head of the investor group funding the Colombian search, requesting another captain in Cartagena.

Upon clearing immigration and customs in Colombia, the remaining "gift" liquor was discretely presented. Then it was time for the crew to spruce up and attend the fancy welcome reception staged for their arrival. The media was there, along with the mayor of Cartagena, the Colombian Navy's bigwigs in full dress uniform and the harbourmaster. The two researchers that John Horton had hired, US Navy Commander John Cryer and Dr. Eugene Lyon, were also on hand. Only the sub's captain was missing. One by one, Helmut introduced the crew, detailing their individual expertise. He had just gotten to his own maritime background when the door burst open and Captain Slaughter staggered in. Annoyed that Helmut was stealing the limelight, he slurred, "Nobody gives a shit who you are. You don't even know how to address a captain properly." Helmut turned and gave the order: "I want him out of here by tomorrow."

The backup captain claimed to have worked in the Beaufort Sea. Ironically, all the ships he listed as his onboard experience were the very ones that Helmut had spent many months on. Curious, Helmut questioned him about various crew members. In the end, the new captain confessed that he had really only been in the Beaufort for one month, leaving because "they had this stupid rule you couldn't have any alcohol." Helmut sent him packing too.

Now the project needed yet another captain, someone with a Masters Foreign Going Unlimited credential. Fortunately, Helmut had a colleague, British retired sea captain L. John Swann, who had those qualifications. Horton agreed. So Helmut called Swann on the sub's sideband radio with the job offer. The problem was that anybody listening could hear their conversation. "We didn't want people to know we were hunting for a treasure ship,"

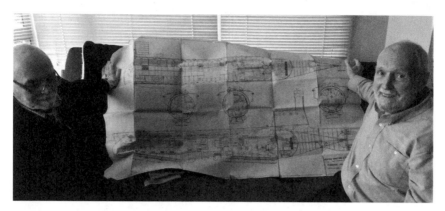

In 2020 Captain John Swann and Helmut Lanziner got together to reminisce, look at old photos, including this blueprint of the submarine, and provide first-hand accounts of *Piccard* stories. Photo by Vickie Jensen

Helmut explains. "Consequently, I had to be cagey about describing what we were doing." Fortunately, Swann agreed to take the job, whatever it was.

When Captain Swann arrived in Colombia, he faced a steep learning curve, since he'd never been on a submarine before. But Commander Fritz Haan, an ex-us submarine commander, came along to provide hands-on training. John Swann soon fell in love with the *Auguste Piccard*. One of the agreements with the government of Colombia was that at least one Colombian submarine officer was to be on board for all dives, thereby gaining deepwater submarine experience while "keeping an eye" on the sub's activities. It was an opportunity that generated mutual respect and support, and even camaraderie.

Despite the crew's repairs while the sub was underway, the *Piccard* still had serious maintenance problems. One particularly memorable incident occurred early on, in 850 ft (260 m) of water. The day began as usual, with the crew up at 4 a.m. and outbound by 6 a.m. Helmut was aboard as well. The sub descended for checks, but came back to the surface because the communication system wasn't working. When diving resumed, smoke filled the engine room. After more repairs, they descended again. At 850 ft, there was an almighty BANG, followed by repeated and equally loud ricochets as a violent

The crew of the *Auguste Piccard* regularly carried out repairs even when the submarine was underway. Even when conditions became seriously dangerous, owner John Horton refused to believe there were problems. Personal collection of Tom Roberts and Helmut Lanziner

spray of water began spurting in under high pressure. Captain Swann immediately ordered closure of the main Barksdale valve (the critical through-hull, high-pressure, directional control valve), followed by "Blow all soft tanks." The sub began an emergency ascent, shooting to the surface. Fortunately, the captain brought the sub under control at about 50 ft (15 m), just in time for a surface check.

Upon investigation, the crew discovered that part of the ballast pump's release valve casing had cracked at depth and blown off, with pieces shooting out like bullets and hitting under-deck piping and compartment bulkheads. Not surprisingly, the crew refused to do any more dives without scheduled time for intense maintenance.

Helmut called John Horton and suggested he come to Cartagena. The two men spent an hour and a half pacing the back deck of the *State Wave*. Helmut recounted the deep-dive problem and explained that there were more daily equipment failures than the crew could repair underway. That meant that shortly the entire search operation would have to shut down. Horton fumed and argued that it couldn't possibly be true since he'd had a crew maintaining the *Auguste Piccard* for years. Helmut offered to show him the failed valve and also mentioned that the new valve replacements were still in their original packages in storage. Horton simply wasn't interested in evidence.

Helmut tried to calm things down, explaining that the crew only needed a couple of weeks of maintenance time to get the sub to minimum operating condition. Horton exploded, shouting, "That's it. I'm going to give her to them. The Colombians can have her." Horton left and flew home. A day or so later, Warren Stearns of Sea Search Armada called Helmut to say that *he* was now the new owner of the *Auguste Piccard*! "I just made a deal with Horton."

John Horton had pulled the plug on his submarine dream.

Warren Stearns quickly put together a company called Chicago Maritime to purchase the submarine. He also arranged for more experienced outside help to come to Cartagena and get the sub diving safely again. At Brian Gorbell's suggestion, Stearns contacted Danny Epp, who took a two-month leave from his job at CanOcean. Danny recalls: "During that time Brian Gorbell was having marital problems and depression so eventually chose not to come back to Cartagena. That's when Stearns asked me to step in as operations manager."

When Danny arrived in Colombia, the propeller shaft seal was leaking so badly that the submarine couldn't safely dive. He immediately grabbed some of the reference and mechanical drawings and then looked at what the crew had been doing with the drive shaft seal. It seemed obvious a part was missing. "So I put my hand down into the bilge, fumbled around in the water and found the part! Then I reassembled the seal with all of the parts and we were back in business. I was the hero for the day! It was one of those rare instances that was a quick, easy fix."

The search begins again

Eventually the additional maintenance neared completion, and the search for the *San José* could finally resume. James McFarlane Sr. and his Canadian company ISE was a shareholder in the project, supplying ROV operators and equipment in the form of its larger *TREC* ROV and smaller *DART*, both equipped with cameras. Although the crew managed to get a considerable amount of video work done with *TREC*, the ROV was eventually lost in heavy seas. Fortunately it had been insured, and the smaller *DART* took over camera work.

Working in the steamy tropical heat, the *Piccard*'s crew was often stripped down to shorts, flipflops and sweat-absorbing bandanas. After the Colombian Special Forces commander complained to Captain Swann that the sub's crew looked like a "bunch of pirates," proper uniforms and insignia were ordered: black shorts and black tee-shirts with insignia, and white "dress" coveralls, also with insignia. Spiffed up in their dress uniform, the crew is as follows: bottom row, left to right: Helmut Lanziner (Cdn/project director), Mike Costin (US/oceanographer), Don Lines (Cdn/communications), Frank Openchowski (US/electrician), Rick Broughton (with beard, US/2nd engineer), George Thomas (Cdn/2nd in command), Mark Corner (Cdn/technician); upper row, left to right: Danny Epp (Cdn/engineer), Captain John Swann (Cdn/captain), "CJ" Savage (US/chief engineer), unidentified Colombian naval officer. Personal collection of Danny Epp, Helmut Lanziner

One late autumn day, the sub was tied up waiting for a pilot when there was a surprise visit by the president of Colombia, along with his heavily armed Special Forces in Zodiacs. Coming aboard, the president shook hands with crew members. When he got to the captain, he asked, "Captain Swann, where is my treasure?" That was followed by a brief inspection to locate any unreported valuables that might have been recovered. There was nothing, and the entourage left as quickly as they had arrived.

At the end of November, Danny Epp's two-month leave of absence was up. He was eager to get back home for Christmas and see his new girlfriend.

Stearns was adamant he couldn't leave because operations had to continue. "He finally agreed to fly my new girlfriend down! She was only 20 years old and had never been south of Disneyland, so coming to Colombia was a big deal." Danny continued working and his girlfriend evidently enjoyed her Christmas adventure because they married eventually and are still together.

Early on in the search, John Horton had recruited Dr. Eugene Lyon, who researched the records of the *San José* in the Archive of the Indies in Seville, Spain. He was joined by retired US Navy Commander John Cryer, who used a technique called reverse navigation, or "re-navigating," in order to predict more accurately where the fabled ship might have sunk, given the prevailing winds and current at the time of the battle. When the search finally got underway over Christmas 1981, it was a smaller crew who searched in the most likely area that the researchers suggested. Oceanographer Mike Costin helped analyze the side-scan sonar, sub-bottom profiler and magnetometer readouts.

They found a convincing target, but was it the actual *San José*? The sub's magnetometer, which senses magnetic fields, registered a spike as they flew the sub over the site. Captain Swann recalls: "We found the target at a significant depth. Then we did a second pass and the side-scan image showed a target with a high end and a lower end, based on the sonar shadow. Of course, the actual ship had been blown up, so it wasn't your typical Spanish wreck as depicted by Walt Disney, but the length was about right." Such a discovery always requires proof and verification, and the contract Sea Search Armada had with the Colombian government stated they were to search for, locate and identify the wreck. So once the target site was established, the crew set about gaining as much information as they could from the site. Helmut insisted on carrying out many runs past the target. In addition to the main target, there were numerous other piles of wood on the bottom in the surrounding area; however, this was the only sizeable target within the primary and secondary search area.

The sub visited the target site a number of times to collect ROV video coverage. One of the earliest visits proved especially eventful. Captain Swann ballasted the sub to stay down on the bottom near the target. Helmut flew the ROV off the main deck of the *Piccard*, taking video footage. Although the ROV got close, it couldn't penetrate the pile of wood that might be the hull. Crew member Mike Costin gazed out the observation portholes in the bow of the sub, making a number of drawings of the wreckage timbers.

Unbeknownst to the crew, during the hours they were on the bottom, the sub slowly settled into the soft seabed sediment, which was many metres deep. When it finally came time to re-surface, the *Piccard* was truly stuck in the mud. Captain Swann pumped out all of the internal ballast to the point where the sub should have started to lift off, but it was still trapped. Realizing what might have happened, the captain sent the chief engineer to check the view from the forward viewing ports where Mike Costin had been sitting earlier. Sure enough, now he couldn't see a thing. The Colombian submarine officer tasked with keeping an eye on the crew was, fortunately, fast asleep in one of the bunks. Anxious not to alarm him or the crew, John Swann tried putting the electric motor on full ahead while putting the steering alternately hard to port and hard to starboard. Simultaneously, he had the crew attempt to rock the sub loose by a coordinated effort of standing by the port bulkhead, then moving quickly to the starboard-side opposite. Finally, the sediment gave up its grip on the sub. Unfortunately, this was followed by a fast and uncontrolled ascent. By flooding all the now-emptied soft ballast tanks, the captain got the sub under control just before reaching the surface. The Colombian officer missed all the excitement.

Ironically, being stuck in the mud yielded an interesting piece of evidence for Sea Search Armada. The *Auguste Piccard* was outfitted with a protective tail cone around the exposed part of the propeller shaft and the Kort nozzle, which also facilitated steering. This cone had an opening that permitted free flooding and inspection of the prop shaft. Afterwards, when Danny Epp did a dive to inspect the tail section of the sub, he discovered pieces of wood that must have lodged in the free-flow section as the submarine sank into the soft mud close to the target.

That wood was precious. It was a real sample from the target site, even if inadvertently obtained. Sea Search Armada had the wood radiocarbon dated and the results showed it was about the right age for a vessel of the *San José*'s era. Captain Swann also recalls that the wood was determined to be of a Spanish species typically used in shipbuilding at the time.

In the meantime, Danny Epp flew back to coordinate the on-going operations of the sub for Chicago Maritime. Then he went on to Galveston to establish a base in the US for the *Piccard* once the Sea Search Armada contract was completed. Brian Gorbell lived close to Galveston and would serve as general manager of Chicago Maritime.

———————————

Sea Search Armada had initially penned a 75–25 partnership, in the company's favour, with the Colombians. Eventually, the deal changed to 50–50, which Warren Stearns agreed to on a handshake. Then, at a second meeting in Bogota, Colombian officials reneged on the 50–50 partnership and demanded 75–25 in their favour. Warren Stearns said, "No deal. That wasn't what we agreed on." Sea Search Armada had invested millions of dollars searching for the wreck of the *San José*, so Stearns jumped on the next plane to Miami and left, telling Captain Swann, "I'm out of here. You guys better demobilize and get out, too."

Captain Swann recalls, "Our agreement with the Colombian government was that upon discovery of the *San José* we would provide them with a position. So we did that." But we also began making plans to leave, based on Warren Stearn's advice. Captain Swann vividly recalls a day of de-mobilizing operations, releasing seabed transponders and recovering equipment. "Upon surfacing, we found ourselves confronting a DAS (Colombia's then notorious Security Service) gunboat with guns trained on the *Piccard*. They announced that the submarine and crew were under arrest. When I asked on what grounds, we were advised only that the submarine and crew had to return to the port of Cartagena. I knew that, as a foreign-flag vessel, all our papers were correctly entered and registered, but there was no way of resisting." Fortunately, as the DAS gunboat escorted the sub back to Cartagena, Captain Swann was advised by a helpful third party that he could seek unchallenged refuge at the Colombian naval base. Consequently, when abeam of the base, he turned the *Piccard* abruptly into the navy's submarine pens. The DAS boat was unable to prevent the quick turn and fortunately did not follow.

The sub's crew had rented a house and apartment close to the Navy yard; now it was obviously time to vacate both. Although technically under arrest, the crew were helped by Colombian naval personnel who had gotten deep-diving experience on board the sub. So crew members were able to obtain fuel for the diesel engines, spare parts, drums of hydraulic fluid and additional food.

The escape plan was that the sub would leave the harbour to "test equipment." However, since the *State Wave* had returned to Morgan City, Louisiana, the real plan was that the sub would now rendezvous with a towboat by the name of *Mr. Cliff* that had been sent to Cartagena to tow them back to Texas.

The sub and its crew made good on their escape on an overcast day early in June 1982. But their adventures were far from over. They encountered the first hurricane of the season, a lengthy storm that battered the sub and its crew. During six-hour bridge watches, Captain Swann and George Thomas, second in command, tied themselves to the conning tower so as not to be swept away by the extremely large seas. Then, still well to the southwest of the Cayman Islands and with the storm still raging, the tug's towline parted at the tug's winch, leaving nearly 600 ft (180 m) of heavy wire towline hanging off the sub's bow. With a heroic effort, the bridle was pulled back on deck and the securing shackle released, freeing the sub from the heavy towline. But the tugboat had had enough and took off.

Left to its own fate, the *Auguste Piccard* submerged below the storm, coming back to the surface every 12 hours to recharge the main batteries. Very slowly, it headed toward Mexico's Yucatan peninsula. Captain Swann made VHF radio contract with the US Coast Guard's seagoing patrol, who were looking to catch drug runners, advising them of the *Auguste Piccard*'s presence. That contact proved fortuitous when the captain suffered an acute kidney stone attack and was taken off the submarine by a USCG rescue helicopter and dropped off at a Cancun hospital. A memorable day or two later, he was able to return to the sub. The hurricane had battered the sub's specially formed Kort nozzle, which was crucial for steering, and now the hydraulic system was leaking badly. As the *Auguste Piccard* finally limped into harbour in Galveston, the sub used up the last of the hydraulic fluid necessary to operate steering. It had been a return trip of epic proportions.

Danny Epp met the sub in Galveston, having set up a small base of operations there. Once docked, the weary crew were released from ship's articles and got off the sub to reconnoitre their next move and maybe even grab a shower. Nobody had been paid for a couple of months; everybody wanted to be done with this adventure.

In truth, no one knew what to do with the submarine. What they did know for certain was that nobody was willing to spend any more money on it.

Once back in Galveston, where it was abandoned for 17 years, the *Auguste Piccard* was stripped of all its instrumentation and became a home to bums and vagrants.
Personal collection of Danny Epp

Sea Search Armada paid some of the owed back wages. Sadly, that was the end of the long, drawn-out adventure for the *Piccard* and her crew. In a strange legal twist, the *Auguste Piccard* would later end up back in John Horton's hands because he had never "technically" sold it, though he made no attempt to move or restore the sub.

When Hurricane Alicia was predicted to hit Galveston in August 1983, Danny Epp had the sub lifted out of the water and put on cradles. "I'm probably one of the last people to have seen the *Piccard* before the hurricane," he says. Sometime later, he returned, just out of curiosity. "The sub had been seized by the sheriff for storage fees. It was in pretty poor condition." The *Auguste Piccard* would sit there for another 16 years after the hurricane. Stripped of all its scientific instrumentation, over the years various bums and vagrants lived in the sub.

Eventually, in 1999, a group of patriotic Swiss brought the *Piccard* back to Switzerland, but it remained a rusting derelict. Finally, the Swiss Museum of Transport in Lucerne (Verkehrshaus der Schweiz) acquired the *Auguste Piccard* in 2005, refurbishing the outside and some of the internal hull. Today the historic sub has its rightful place as an impressive exhibit alongside the planes, trains, funiculars, steamboats and balloon gondolas featured in the museum's collection.

It's worth adding that the hunt for the *San José* continued. Horton had the right idea, just maybe not quite the right technology or timing for finding the fabled treasure ship. On November 27, 2015, an international partnership of Maritime Archaeology Consultants (MAC), Switzerland AG, and the Colombian government retained the services of Woods Hole Oceanographic Institution (WHOI). Woods Hole used the AUV *REMUS 6000* to conduct long-duration surveys of the area. *REMUS* also confirmed early side-scan sonar images of the

Guy Immega visited the *Auguste Piccard* in Lucerne after the Verkehrshaus der Schweiz (Swiss Museum of Transport) acquired the sub and refurbished the outside and some of the internal hull. Personal collection of Guy Immega

wreck, going in just 30 ft (9 m) above and taking photos of decorative dolphins on the sunken ship's cannons that helped clinch the identification.

However, the fate of the *San José*'s treasure continues to be in dispute. Sea Search Armada is still actively suing for half the value of the *San José* in an ongoing legal battle that has now lasted more than 35 years. Several nations are also suing for part or all of the treasure, including Spain, which claims ownership of all sunken Spanish ships. That part of the story is far from over.

CHAPTER 12

Inuktun: Smaller is better

1989–present day

As a six-year-old, Al Robinson built model airplanes and dreamed of flying De Havilland Vampire jets. "But in my early teens they discovered my eyesight wasn't good enough to fly those jets, so my mother and I decided that I should become an aeronautical engineer. However, in grade 11, I dropped out of school so I could build hot rods and custom cars and chase girls. My only claim to air fame is that one of my designs was displayed at the Smithsonian Air and Space Museum. Airplanes have remained a significant love all my life, but I ended up going the other way—one and a half thousand feet *down*, underwater."

Like so many subsea trailblazers, Al Robinson started out working at HYCO. "I was a truck driver, a carpenter and an auto-body worker when I arrived there. I had no knowledge of underwater stuff. I couldn't even swim because I'm one of the few people with negative buoyancy. But HYCO hired me as an assistant, and then they fired my boss and asked me to take over. That's how I became chief hydraulics technician." (See Chapter 4.)

In short order, Al Robinson picked up not only the technology but the company's can-do-anything attitude. "I came in the early days of the company, when HYCO had just gotten a contract with Vickers Oceanics, so they were scrambling to accomplish things. It was hectic, but it was also incredible. These were people like I had never known before. Right from the very beginning, I loved the work."

When HYCO closed, in 1979, Al moved to International Submarine Engineering and somewhere in there designed a manipulator for the Woods Hole submersible *Alvin*. After he left ISE, his first wife, Melinda, and their son Scott worked there. "Melinda was very talented at electronics assembly, especially wiring." Today Scott Robinson is a mechanical designer at the Nanaimo-based company Inuktun, which Al and Terry Knight started in 1989, and Scott's sons Dean and Kevin are technicians there, as well.

Al always loved the challenge of technology, no matter what company he was working for. After leaving ISE he worked for Jack Wilson, head of Robotic Systems International in Victoria. Originally affiliated with International Submarine Engineering, RSI became independent in 1983. Jack was considered "a visionary leader with a great mind and a nice guy all around." At its largest, there were maybe 35 on staff. Ken Soles was the main mechanical guy, and Al Robinson handled design. Chris Roper looked after sales and business development. Norm Keevil was a mechanical engineer who took a six-week contract with RSI and stayed. David Lokhorst ran RSI's special projects division for several years.

One of the most unusual RSI challenges that Al took on was creating a bunch of mechanical spiders for *Runaway*, a 1984 sci-fi action movie about rogue robots. In 1986 RSI played a more traditional role in the salvage of the *Challenger* space shuttle, which had exploded soon after launch, killing all seven astronauts aboard. Given a "right now" deadline, RSI delivered a seven-function and a five-function KODIAK manipulator. Chris Roper was commended for assisting intern ROV pilots with the operation of the manipulators and the subsequent recovery of one of the two hefty solid rocket boosters (SRBs) and associated hardware. Although RSI didn't last terribly long as an independent venture, it's a good example of numerous small subsea tech companies that got their start in conjunction with larger entities such as HYCO, ISE and Can-Dive.

Terry Knight and Al Robinson go further back to HYCO's early years. "We had our little shop facilities side by side," Terry recalls. "He was the hydraulics wizard and I was the sparky. As a result, we spent a lot of time working together, transposing each other's drawings so I could learn about hydraulics and he picked up on electronics and electrical. Al's a smart man, but also a real character. He was never a guy to go out and make money, but he *loved* technology. And he became an extremely talented mechanical designer."

Inuktun partners Al Robinson and Terry Knight examine the small tracked vehicle Al initially designed for RCMP bomb disposal. Capable of changing shape and climbing stairs, he dubbed it the *Variable Geometry Tracked Vehicle* (*VGTV*). Personal collection of Terry Knight

Years passed, and Terry and his second wife came back to Vancouver Island, then took off to work government jobs in Inuvik. During that time, Terry kept in touch with Al, even inviting him to come up for a visit. "When I'd have to go places like Tuktoyaktuk, Al came with me. It was freezing cold, so there was nothing to do after work but sit around with a bottle of whiskey and talk. That's usually where brilliant ideas come from, and that's exactly what happened. Al talked a lot about an ROV project he was working on in his spare time called *SeaMor*."

Eventually Terry and his wife moved back to BC and settled in Nanaimo. Al was still working at RSI, but often he came up to Nanaimo for visits, bringing along an early prototype of *SeaMor*. "We'd take that ROV down to the local marina where I kept my sailboat and we'd play with that thing all weekend, trying to figure out how to make it better, how to fix things," Terry says. "We kept burning out thruster motors. They were actually sewing machine motors and just happened to be exactly the same motor that my wife had in her sewing machine, so I'd 'borrow' it, take it back to the marina and we'd play some more." Come Monday morning, Terry would phone a supplier in Vancouver and buy a couple of sewing machine motors for Al's next visit.

"Then we got carried away with ourselves and started talking about maybe starting a business and manufacturing these things. We could sell them all over the world. People would be dying for them!" Back then, not many companies were doing that, and the ROVs Al and Terry were working on were some of the smallest on the market. "Well, neither one of us was quite the genius we thought we were, but we stumbled along and finally started selling some."

"We called our company Inuktun Services because I'd been up north," Terry continues. "We figured nobody would have that name. Inuktun means 'in the way of man' or 'like a man' in the Inuktitut language and I guess that actually describes what our vehicles do—they look and inspect in areas where a person can't go. Anyway, starting in 1989 the business took off from there." That explains how two friends, Terry Knight and Al Robinson, got to indulge their techie side and play around with inventing and building a small, remotely operated underwater vehicle. Al adds, "When we started Inuktun we were mainly concerned with underwater vehicles, but soon became involved with other systems including crawlers for things like inspecting sewers and other pipe systems. The first crawlers we created were *MiniTracs*, which became a major product for the company."

Al and his first wife had already split up, and he was thinking of starting a company in the valley north of Hatzik Lake, near the Fraser River, where he could just hide away. "But when Terry and I decided to start a company together, and he already had family in Nanaimo, so I brought my mobile home over from the mainland and put it back in the trees on a piece of property that Terry's dad had bought. I thought I'd live the rest of my life there all by myself, working away at the company right close by."

The company's very first contract was with Phil Nuytten, modifying one of Can-Dive's vehicles to do some pipe inspection. "It was just a small contract but it got us going," Al says. "Soon we had enough work that we hired six or eight employees." Terry adds, "We built an even smaller ROV that we called *Scallop*, but we had to shelve it for a couple of years while we worked on *SeaMor*. That business grew quite well, and we also built a set of strong, tiny crawler tracks, almost like miniature bulldozer tracks, that could bolt onto the bottom of *SeaMor*. Or you could just put a camera on it for sewer inspection. We came up with a number of neat widgets."

Both recall the day in 1996 that changed their company plans dramatically. "The Department of Highways was re-doing the Vancouver Island highway and told us they were expropriating the property. They wanted to cut our property on the diagonal, leaving us the half that fronted onto Cedar Road, and only pay us for the cheap back part. But the promised cheque kept getting delayed," Al recalls. "We even hired some lawyers to fight it, but we just kept hitting a brick wall. Finally, they agreed to buy the whole property, but by then our bank had cancelled our line of credit and wouldn't even honour the cheques we'd already written. That's when we realized we'd have to shut the business down. We called all our employees together and told them, 'This is the end of the road for the company. It's all over.' All but one of them said, 'We don't believe you. You'll sort this out and get going again. And when you do, we'll be back to work.'" That vote of confidence included Colin Dobell, who would become Inuktun's president and CEO. But back then, he'd been an employee for less than a week. When Al handed him his layoff notice, he said, "I want to work with you guys. So let me know when you're back up and running, and I'll be there."

"That very afternoon, when we were trying to shut down the business, the damn telephone kept ringing," Al says. "And one of the calls was a confirmation for the biggest contract we'd ever had. There were several more things like that, so we dusted ourselves off and got back at it, moving the shop to an existing facility under Ming's Store, a small grocery in Cedar, a community a few miles southeast of Nanaimo. Colin and most of the others came back and we got busy building the biggest vehicle we'd ever done." Pacific Gas and Electric needed to inspect old gas pipes in California to see if the company could still use them. "So we built lots of tracks because the ROV in that system had three pairs of mini tracks in order to carry a camera and ultrasonic testing equipment through those gas pipes."

Back then, Inuktun didn't have the full capability to handle the electrical system for the whole vehicle so sub-contracted that to a company in Vancouver. Al recalls they had a tight deadline with severe penalties for every day over the deadline. "When Terry phoned the Vancouver company the day before they were supposed to deliver, the guy admitted, 'Well, we haven't started on it yet.'"

Now Inuktun was in real trouble. "Terry talked to the National Research Council, who put us in touch with a company in Chilliwack who agreed to come in and do that part of the project. And Jim McFarlane at ISE sent one of his top people over. When it was finally completed, Roy Coles and I drove the vehicle to the company in California that was contracted to do the actual

Paul Prunianu working on the large ROV for Pacific Gas and Electric. Personal collection of Al Robinson

instrumentation for taking the ultrasonic readings. We still didn't have our part of the vehicle quite done, but we finished it in their shop, with people like Colin and others working insane hours. Then we did the testing and delivered it—but it was late."

At the acceptance meeting, the chief engineer from Pacific Gas and Electric stated, "Yes, you were late, but we've seen what the problems were, and we've watched how the Inuktun people have performed, so we're eliminating the late penalty." Then he added, "And we're not happy with the inspection part of the system, so we're giving Inuktun a new contract to do that part of it as well." When all that was over, in 1996, Terry and Al decided to give Colin part of the company and make him president—a good decision.

Sometimes it takes years for a vehicle idea, such as one of Inuktun's specialized miniature ROVs, to come to fruition. For example, when Al was working with RSI, the RCMP bomb squad people had a remote-control bomb disposal vehicle with problems. They had asked Al for advice and followed up on a couple of his suggestions. But Al recalls, "When I left there, I knew I could produce a better vehicle than that." So he went home and played with a design, building parts of the mechanism with his old Meccano set. "I designed a vehicle that could stand up and lie down and go up stairs. It could also stand up tall and skinny in order to turn in hallways. It had a manipulator and video cameras."

Not long afterward, when Inuktun got underway, Terry arranged for the two of them to talk to the RCMP bomb squad again. "We didn't even have any good drawings or much of anything to show them," Al says, "so my son Scott, who had joined the company, and I built a quarter-scale working model of what I had envisioned. Scott also built a set of stairs with a doorway for the robot to go up and down. The RCMP were very impressed and wanted us to build a full-sized working model to send to Ottawa for evaluation. But we didn't have the money for that. However, on our own we eventually developed a robot for explosive ordnance disposal."

In the meantime, Chris Roper joined Imagenex in sales. He and Al had both worked for RSI. Now he phoned Al to say he had just sold a *SeaMor* ROV to a nuclear power plant in Peach Bottom, in Pennsylvania. Al's immediate reaction was, "Oh, no, no, no. You can't do that. It's not made out of the right materials for that kind of toxic environment. And I don't know enough about nuclear stuff to build one out of the right material." But Chris explained, "They recognize that, but they're willing to buy *SeaMor* and use it as much as they can because they like it and it's affordable."

Scott Robinson with a stainless-steel version of *SeaMor* for a Japanese nuclear company. Personal collection of Al Robinson

Soon after that, Terry and Al flew to the eastern US, visiting Peach Bottom and another nuclear plant to see what the nuclear world was about. "We stood on this nice clean floor while right underneath us was a reactor producing several million watts. That trip got us into the nuclear business, and we've done a lot of work in that area ever since. In fact, we sent a custom snake-like crawler that went down into the bottom of the Fukushima Daiichi Nuclear Power Plant reactor after its meltdown following the 2011 tsunami."

"Actually, there are two kinds of nuclear people—there are nuclear *power* people and nuclear *weapons* people," Al clarifies. "Nuclear weapons people were very active during the Cold War and now they've got some horrendous messes to clean up. But first, they need to inspect those dump sites and evaluate the state of those containers. So we've been to Hanford on the Columbia River in eastern Washington with a lot of our vehicles working there to look at their waste areas. And when 9/11 happened in New York City, our vehicles were the only ones successfully operating in the debris at Ground Zero—and the only ones that found human remains there. I remember watching news coverage on CNN, seeing somebody put something down a pipe to go way down and check the foundations under the debris. And the announcer said, 'It's called an Inuktun and it comes from Canada.'"

"Our involvement with 9/11 came about because some search and rescue type people in the States already had our equipment," Al continues. "But what they had wasn't quite right for this terrible situation. So they called us right away and said, 'Can you do some vehicles for us that will fit in backpacks so they can be carried into the debris and that are battery-operated so they can operate without outside power?' We had never done anything like that before. And then they added, 'And can you do it by this afternoon?' Terry talked to Industrial Battery and Supply, a battery supplier in Nanaimo, about getting some special batteries quickly and they stated, 'No, we can't do that.' But when they heard it was for 9/11, they said, 'The batteries will be there this afternoon.'"

Similarly, Al rounded up backpacks, cameras and a camera case for carrying cameras, but the case needed more straps and pockets sewn on. So he went to a store that did garment repair and alterations in Nanaimo's Rutherford Mall (now Nanaimo North Shopping Centre) and the staff agreed they could do that. When Al told her what it was for, she said, "It'll be done this afternoon and there will be no charge."

"It was just amazing how everyone was more than willing to help out," Al recalls.

The right time and the right technology

During the 1960s the Space Race surged between the US and the Soviet Union. It was also a time of technological advances such as miniaturization, lightweight metals and emphasis on communication. Whether it was for space or underwater exploration, innovations seemed to come together at the same time to create an impetus for subsea work.

"It has to do with the technology arriving at the point where a particular thing is available or doable," Al says. "For example, in 1850, people were already trying to fly airplanes, but back then the only power available was a great huge steam engine. By 1900, there were small gasoline engines; as a result, in 1903, a lot of people were ready to fly. Regardless of who flew first, it really happened because technology had gotten to the point where it could be done.

"With the underwater stuff, the metal we needed for hulls couldn't have been produced in 1850. Back then it would have been a great huge cast iron thing that was simply not workable. Whereas by the 1950s we were getting various steels and types of aluminum that had the strength and viability to do the job. In addition, after the Second World War there was an explosion in the quest for scientific knowledge that drove exploration in the ocean and in space. A lot of those accomplishments came out of the competition between the US and the Soviet Union. Many achievements on both sides were done to gain power over the other."

Al notes that miniaturization is an important part of what Inuktun does. His son Scott is working on drawings for a mini-track vehicle. And there's a micro-track one and now a nano-track one for Hanford. These will be delivered by a bigger vehicle with a smaller one that comes off it and can even go into grooves in a floor under tanks. "You have to keep pushing the limits in this business. And you need money to develop your ideas." Al shows a picture of *VideoRay*, noting, "We built that ROV here as *Scallop*, but we didn't have the money to put it into mass production, so we sold the technology to an American, Scott Bentley. It was renamed *VideoRay* and apparently he's sold over 3,000 of them now." Eventually, Inuktun also sold the *SeaMor* ROV to a company that still exists in Nanaimo as SEAMOR Marine.

Having outgrown the facility under Ming's Store, Inuktun made a wise move to an industrial area in Nanaimo. But as with all maritime industries, the business can be feast or famine. "Fortunately, we have incredible people here, and the few times we didn't have the money to make the next payroll, people accepted half a paycheque to keep the company going. That's saying a lot!"

Al shakes his head. "Ironically, we sell almost nothing in Canada, but that's normal. And few people here in Nanaimo know that we exist or what we do. We've done a lot of things for the US military, but I don't think we've ever had a contract for anything from the Canadian military. We've done proposals, but they seem to think we're just locals so couldn't produce anything good."

The US Army Corps of Engineers has no such hesitation.

In 2003 Al Robinson and *SeaMor* made the cover of *International Ocean Systems* magazine. Personal collection of Al Robinson

"They contacted us when they wanted a remote-control vehicle to send through cross-border tunnels rather than sending a person. So we adapted a pipe inspection vehicle, and they bought several. Then they wanted one to use when there's a tunnel that ends on private property; they wanted to be able to bore a hole and have something to lower down through that hole and then reassemble itself like a transformer. I told them we could do that, so they gave us a development contract and we built their vehicle. They've used them in cross-border tunnel investigations at US–Mexico borders and also overseas in the Middle East."

Al once asked US military personnel about outsourcing: "You're the biggest industrial country in the world. Why don't you have the work done at home?" He still remembers the reply: "Because nobody will do this kind of thing; nobody will step outside of the box."

Although he is now retired, an interview for *Deep, Dark & Dangerous* brought Al back to Inuktun for a visit in son Scott's office. Over the years, Al has also been grateful for help from his family. His first wife, Melinda, remarried Roy Coles and both worked at Inuktun for several years. Melinda did electronic assembly and built small video cameras. Al's daughter Pat often contributed ideas and provided PR materials for the company. When Al retired, Scott took over his job as mechanical designer. Scott's two sons also work at Inuktun. Photo by Vickie Jensen

Former Inuktun president Colin Dobell explains that the company no longer does underwater work. Inuktun sold that technology to a company in the US and now focuses on confined space crawlers, cameras and inspection equipment. "We do produce some pretty remarkable vehicles for pretty unusual situations. For example, Starbucks bought our cameras to inspect coffee beans on a conveyor. But our most compelling stuff is top secret—that's because our big-name clients require that we sign non-disclosure agreements. They don't want the public to know what they're using the technology for. It certainly keeps the work interesting!"

Terry Knight worked at Inuktun until the day he quit in 2005. "I went to work one morning, sat in my office for a while, then picked up my keys and my cell phone, and dropped them on Colin's desk. I told him, 'I'm not having fun anymore.' Strangely, I didn't realize I was gonna do that, but I did." Al stayed on quite a bit longer, only retiring in 2014, at age 77.

Looking back on Inuktun's early decades, Terry adds, "It was sure good for both of us during those years."

Al puts it this way; "There are so many stories on the coast. I have always been absolutely amazed and thrilled to be just a little part of it."

Recent Company Update

In February 2019 Inuktun was purchased by Eddyfi Technologies and renamed Eddyfi Robotics Inc. The Quebec-based parent company produces advanced Non-Destructive Testing (NDT) solutions for a broad market. The Nanaimo facility is now the Eddyfi Technologies' Center of Excellence for Robotics, but it is still the home base for developing and manufacturing remote visual inspection (RVI) and robotic crawler solutions for complex and challenging situations worldwide. At the time of the sale, Inuktun had 65–70 employees and offices in the US, Europe, Singapore and China. Not bad for a company that started out as a retirement project!

CHAPTER 13

Subsea science

S cience is perhaps the most respectable version of human curiosity. Those driven by curiosity push to glimpse the unknown, to figure out how a thing works, to document where it lives, or to track vehicles using acoustics. This chapter profiles three BC scientists and one private contractor, their underwater work and how it has shaped their lives.

Dr. Verena Tunnicliffe, University of Victoria

In the 1960s, every curious kid dreamed of becoming an astronaut. Verena Tunnicliffe was no exception. Fortunately for BC, she built a notable career as a remarkable marine biologist instead. She's only recently retired from a joint appointment in the University of Victoria's Biology Department and in its School of Earth and Ocean Sciences. She's also held a Canada Research Chair in Deep Ocean Research. The Tunnicliffe Lab studies the nature and functions of deep-sea communities. She's made hundreds of dives with scuba equipment, manned submersibles and ROVs in order to explore BC's fjords, seamounts, hot vents and subsea volcanoes. Her research in hydrothermal systems helped establish Canada's first Endeavour Hot Vents Marine Protected Area (MPA).

Verena grew up in Deep River, Ontario, in a family environment teaming with science and marine interests. Her father was a nuclear physicist, basically building a nuclear reactor using a slide rule. Her brother became the fleet oceanography officer in Canada. She attended Yale for graduate work, intending to study the impact of waves on coral reefs, but an engineer on her PhD committee steered her into the study of fluid dynamics. "I learned the difference between a biologist and an engineer—a biologist works to the

Dr. Verena Tunnicliffe examines a sea lion skull from the bottom of Douglas Channel, retrieved by the scientific ROV *ROPOS*. Personal collection of Verena Tunnicliffe

left of the decimal point and an engineer works to the right of the decimal point!" she laughs.

Graduating with a PhD in ecology, she started her job search. But she was a woman and a non-American, so didn't get a single job offer in the US. Finally, she applied for a two-year post-doc at the Institute of Ocean Sciences, run by the Department of Fisheries and Oceans (DFO) in Victoria, studying the anatomy of a worm found in polluted environments. "It was the least romantic bit of scientific research you could imagine, but I was desperate. Victoria felt like a remote outpost when I arrived. I was 27 and sure I'd only stay for a couple of years."

When Verena was done with her daily worm work, she used to wander around the Institute of Ocean Sciences (IOS) in nearby Sidney. *Pisces IV* was in pieces on the dock, and she wondered what was going on. "I was a female in a skirt, but the guys there got it that I was interested. So they offered to take me up to Jervis Inlet, the deepest fjord on the BC coast, for a test dive once they got the submersible put back together. That dive down to 732 metres changed my life! Growing up, I had always wanted to be an astronaut. Now I wanted to be an aquanaut."

Verena went back and told her post-doc supervisor that she was done with worms. When she spotted *Pisces IV* as a young scientist, the submersible was practically unused. Eventually, she would clock 120 dives in it. In the early '80s, *Pisces IV* was unbelievably basic. It had a torpedo arm that was a single, clunky five-function manipulator, going back to the days when *Pisces* recovered torpedoes. There was no camera, no nothing. Verna had to crawl across the pilot to take a photo. "Back then, the only other female scientist working at IOS was Ann Gargett, who built a way to measure turbulence. She had done some dives before me with *Pisces IV*, but I was the first woman to go offshore with the submersible."

After landing a faculty position at the University of Victoria, one of Verena's first expeditions was documenting life on Cobb Seamount, an underwater volcano that's 500 km (310 mi) west of Grays Harbor in Washington State. "One day I was staring at a chart with the depth marked in fathoms and noticed that the Cobb Seamount rises 2,743 m (8,999 ft) up to a flat pinnacle that's only 18 m (59 ft) below sea level, so it's a real navigation hazard in high seas. It also extends into the upper ocean region where light penetrates. Not surprisingly, it supports a dense ocean ecosystem. I got fascinated with it and started to read whatever I could find about it. Somebody had tried to set up a weather warning system on it. And earlier, the US had tried to set up a listening post there, complete with hydrophones, to detect Soviet subs. So I decided I'd go take a closer look with *Pisces*.

"Somehow, the US found out that I was going to do that, and I got a call from the US Navy saying I would be ill-advised to continue with that plan. I replied that the Cobb Seamount was in international waters so I was pretty sure I could go ahead." At the same time, the University of Washington's ship *Thomas G. Thomson* was looking for hot vents in the general area, so Verena contacted U Wash geologists John Delaney and Paul Johnson. They had some unpublished information about Cobb that they shared with her, so she reciprocated and asked if they wanted to go down with her in *Pisces*. "So that's what we did."

In 1982 the group did six submersible dives on the Cobb Seamount. At that time *Pisces* wasn't diving to its full rated depth and the mothership was still using the Loran C navigation system, a GPS predecessor that determined position from radio signals broadcast from land. Verena describes it as "awful."

After she left, the ship dredged on the Endeavour Segment of the larger Juan de Fuca Ridge southwest of Vancouver Island. It's a seafloor spreading centre and now is noted for being full of hydrothermal vents and black smokers. But back then it was pretty much unknown territory. Verena recalls getting a call from one of the team saying, "I've got all this stringy stuff attached to vent rocks that we dredged. Do you want it?" She did indeed. "That was the first discovery of hot-vent animals on the Ridge."

During her career, Verena has gotten to meet some fantastic people, including Anatoly Sagalevich from Russia. "He's one of the bravest people I know and really moved the field of subsea research forward with *Pisces* and *Mir* submersibles. I first met him at a conference in Estonia in 2012, where he greeted me with his booming voice and heavy Russian accent, 'Verena Tunnicliffe, you stole my sub!' But he said it with a grin. And he was right!" The Russians had ordered *Pisces IV* from HYCO, but when the US government got wind of the transaction, they stopped the sale (see Chapter 4). As a result, the Canadian government had to buy the submersible in order to bail out the company. Luckily for Verena, *Pisces IV* was eventually sent over to the Institute of Ocean Science in Victoria.

Pisces IV had always been capable of diving to 2,000 m (6,560 ft) but had never been certified for that. Once that was accomplished, Verena made an expedition to the vents on Juan de Fuca Ridge. "John Delaney, Paul Johnson and NOAA were doing bathymetry, mapping the contours of the ocean floor. They were also studying the Axial Seamount, the youngest volcano in the Juan de Fuca Ridge, so I really wanted to get down there." Dick Chase, from UBC, joined the team, as did Steve Hammond from NOAA. It was a great collaboration. "We dove on Axial Seamount with only a one-page Xeroxed copy of seafloor geological structures for a chart. The results were amazing—the first discovery of hot vents on the Juan de Fuca Ridge (at 1,570 metres) and animal species never seen before."

In 1984 Verena got to dive in Woods Hole's submersible *Alvin* to the Endeavour Segment, where Delaney and Johnson had dredged the vent samples in Canadian waters. "We found a big field of hydrothermal vents with black smokers and more new animals. That year, I also fielded an expedition to Explorer Ridge with another set of discoveries." And in 1986, Verena led a trip with NOAA and *Pisces IV* back to Axial Seamount, where they found another large vent field. Geologists at the BC Department of Energy, Mines and Resources were instrumental in supporting this expedition, Verena says, writing letters to encourage funding. In 2003 the Endeavour Hydrothermal Vents Marine Protected Area was designated to ensure protection of the vents and the unique ecosystems associated with them, a hugely important step forward.

In 1986 the Institute of Ocean Sciences lost its subsea capability when *Pisces IV* was given to the Canadian navy for training purposes. Funding is always difficult for oceanography and subsea research, with three basic sources of money—military/defence spending, private industry, and academia. The US seemed to have better integration of their funding, she adds. And certainly Canada has smaller pots of dollars. Fortunately, the federal government at that time had an Unsolicited Proposals Program. Building on the fact that Vancouver and Victoria were the focus of all the science, the interest and the robotics that were happening, International Submarine Engineering (ISE) proposed a contract to build a science-class ROV to be operated by the Institute of Ocean Sciences; the proposal was accepted, and the end result was ROPOS (see Chapter 8). "I think that ROV was another idea that pushed the subsea world forward."

"In the beginning, there was a lot of competition to conduct this kind of research," Verena notes. Woods Hole had *Alvin* and a prototype ROV (*Jason Jr*), but no one had the capability to go very deep. Then the Institute of Ocean Sciences took delivery of a 5,000-m (16,400-ft) cable for ROPOS and things looked good. "Of course, the very first dive never got to full depth because the feedback began shorting out. It turns out we needed a cable that had flexibility to deal with wave surge. Without it, the stress broke the fibres inside the cable over and over. It took two more years to build a cable that worked. That part was quite depressing, but those first few hours with ROPOS were so exciting! What was really amazing was that scientists, technicians, and the ship's captain could all sit together watching the video monitor at the same time.

"Initially, ROPOS was really almost a prototype, but the ROV certainly evolved as we went along. Meanwhile, the guys in Woods Hole were seeing what we were doing and were making their own improvements. We were just all leapfrogging each other, along with the other global research scientists. First there was a new *Jason*, then ROPOS was improved, then *Jason Jr.* And then the Japanese entered the picture with their *Kaiko*. Then the Germans came online with *Quest*, then the French with *Viktor*, and it kept on like that. In the early days, the investment by NOAA was critical. They realized that the only way they could keep their vents program going was to fund ROPOS expeditions."

Then, in 1996, just as ROPOS was proving its value, the ROV was lost off the West Coast during a severe storm that also partially disabled the research vessel. Verena recalls: "I thought that my career was over. But losing the ROV was maybe the best thing that could have happened, because we had insurance!" That unexpected windfall allowed everyone—scientists, engineers, and students—to give input on a new design. The re-configured ROPOS II arrived in 1997 and eventually was the first to have an HD video camera, so no more fuzzy video. Verena says, "I was in tears the first time I saw the amazing views it took."

Around this time the government released ROPOS II for the private sector to operate. Thus a group of Canadian scientists, initially led by Steve Scott at University of Toronto, established the Canadian Scientific Submersible Facility (CSSF). Verena's husband, John Garrett, was instrumental in setting it up as an independent non-profit company. Reliance on grant funding slowly decreased as global work picked up. An additional spinoff of this global work is that ROPOS II has gained a reputation as the world's most capable scientific ROV.

Enter *VENUS* and cabled observatories

But Verena Tunnicliffe wasn't done with dreaming. On offshore expeditions, scientists discussed how they might connect scientific instruments on the volcanoes and hot vents to shore so that they could monitor changes continuously. In the late 1990s John Delaney shared many ideas as he looked to study the entire Juan de Fuca tectonic plate. However, US funding faltered, so the first opportunities to build a test of that concept came in Canada through the newly established Canadian Foundation for Innovation. Verena assembled a group of interested scientists, who wrote a successful proposal.

Thus *VENUS*, the acronym for Victoria Experimental Network Under the Sea, was launched, designed by a team led by Adrian Round. Critical innovation was provided by OceanWorks Inc. in Vancouver, as they built the power and communications nodes for seafloor deployment.

In 2006 the *VENUS* coastal observatory, located in the Salish Sea, became the world's first cabled seafloor observatory with open, online access—the idea being to test the technology in sheltered and tectonically quiet waters before deploying it in the open ocean. Operations began in early 2006 with a cabled array installed in Saanich Inlet. A year later the *VENUS* observatory expanded into the tidally more active Strait of Georgia, with a special extension into the Fraser River delta. The data centre built at the University of Victoria also served as a prototype for the next stage.

In 2009 the world's largest cabled ocean observatory, called *NEPTUNE*, was installed in the northeast Pacific Ocean, with cables across the Juan de Fuca plate to the Endeavour Hot Vents MPA, among other locations. Today, both observatories are operated by Ocean Networks Canada (ONC) at the University of Victoria. *NEPTUNE*'s greatly extended network spans a range of ocean environments, depths and locations. In 2014, the US National Science Foundation's Ocean Observing Initiative also built a cabled line that extended out to Axial Seamount. Thus the northeast Pacific Ocean is one of the best-monitored ocean regions in the world.

Keith Shepherd and the Canadian Scientific Submersible Facility (CSSF)

In 1980 Keith Shepherd had a brand-new diploma from Camosun College certifying him as an electronics technologist. He landed a job with the Department of National Defence (DND), and then moved to the Institute of Ocean Science in Sidney. IOS was operating *Pisces IV* at the time, and Keith was hired as a pilot/tech. That's where he first met Verena Tunnicliffe.

But piloting was just one of many duties with the submersible. "We had a team of six who shared responsibilities, repair, and maintenance of mechanical, electrical, hydraulic and navigation systems." Keith got to know *Pisces IV* intimately, designing and rebuilding the submersible's entire interior electrical system. He travelled with *Pisces IV*, diving on the East Coast, in the Arctic and on the West Coast, then switched to *Pisces III* for a US scientific research project for NOAA in the Gulf of Mexico. Keith especially remembers the first dive with *Pisces IV* at Axial Seamount and the excitement of that first discovery of hot vents on the Juan de Fuca Ridge, along with brand-new animal species.

Then the DFO cut back on funding for the Canadian support ship *Pandora II*, but continued funding *Pisces IV*, which needed a support ship to conduct research. Instead, the DFO bought an ROV from International Submarine Engineering. *ROPOS* arrived in a bunch of boxes to a mixed welcome. "At that point we had

Keith Shepherd working on *ROPOS*. Personal collection of Keith Shepherd

Pisces IV running like clockwork. Now we were seconded to put *ROPOS* together and operate it with no budget. It was super frustrating."

In 1996, three concerned scientists—Dr. Steve Scott, Dr. Kim Juniper and Dr. Larry Mayer—formed the Canadian Scientific Submersible Facility (CSSF). As scientists who used both *Pisces IV* and *ROPOS* in their research, they were troubled when they heard that the Canadian government was going to sell both the submersible and the ROV. That would have been the end of any Canadian deepwater research.

At about the same time, Keith Shepherd's position with IOS was declared redundant, so he jumped ship, switching from government work to the private sector. He set up his own company, Highland Technologies Inc., to operate and manage the new CSSF facility. Eventually another Keith, this one Keith Tamburi, moved to Vancouver Island to work full-time with Highland

Technologies. The CSSF leased ROPOS from the Institute of Ocean Sciences and operated it on a not-for-profit basis, slowly building a following in the scientific community primarily by hard work. "Our budget from the Institute of Ocean Sciences was zero, so our destiny was in our own hands. We charged a day rate, simply a fee for work."

Keith Shepherd has worked with Verena Tunnicliffe for over 35 years, first with *Pisces IV* and then with ROPOS. "Verena and others really stuck with us during the early years when the ROV wasn't functioning well, because they saw how hard we were trying to make ROPOS work. It was a great collaboration. In those early days, Keith Tamburi and I were offshore most of the summers, so I saw very little of my kids."

Keith Shepherd explains, "From the beginning, our goal was to make getting the science easy." That meant that when a scientist would show up with a research problem but didn't know the technical details of how to get it done, the two Keiths would guide them through the process. Sometimes it was "Let's try this and this" or "This won't work but what if we try this?" Together they'd figure out a workable procedure. That approach was really appreciated throughout the scientific community and earned CSSF a loyal following.

Then came a particularly painful experience early in October 1996, when ROPOS was lost while on a trip to the Endeavour Ridge area. "We always keep a close eye on the weather," Keith Shepherd states, "and knew that a couple of systems were coming in by the next day, so we dove the ROV overnight, expecting to recover at breakfast. But the two systems came together to form what is called a weather bomb, and within two hours the winds had gone from 12–15 knots (22–28 khp) to over 80 (148). We still had ROPOS in the water and couldn't recover the vehicle in those conditions—it's a two-hour process. The support ship was trying to hold station, but then lost its bow thruster and one stern thruster, then slid over the ROV and cut the umbilical cord. That was the last we saw of ROPOS."

When the insurance money came through and discussions began about the design of a successor, the biggest upgrade needed was a different type of connector. "Correcting the connector problem was a big change; our connector reliability went from poor to excellent." Another major upgrade was having all components connected to a telemetry system—that's the computer control system for operating the vehicle. It also includes the technology and equipment to have the surface and subsea computers communicate through the ROV's umbilical cable.

Over the next years, the CSSF got more global work with ROPOS II, helping to spread the work schedule around the calendar. But the quest for work is always ongoing. Keith Shepherd says, "The irony of a not-for-profit facility is that if you make money it's bad because a lot of scientists don't want you to get rich off their grant. But if you don't make money, you die. So our existence is always on that razor's edge. Any money we make, we put back into the vehicle. As a result, it's now a state-of-the-art piece of research equipment. In fact, a lot of other organizations have changed their systems to copy ours."

In 2004 the CSSF fully took over ownership of ROPOS II and rebuilt it completely. Keith's company did the design and rebuild in-house. A year later, the rebuild was followed by several new systems: a side-launch LARS (launch and recover system) and a new telemetry system.

Neptune comes online

John Delaney, of the University of Washington, and Keith Shepherd are good friends, going back to *Pisces IV* days. When the *NEPTUNE* cabled observatory was set up in 2007, Highland Technologies did a lot of the initial installation work. The work included providing quality control of cables, connectors, instruments, and the instrument frames loaded with sensors, cameras, control systems, and so on. This meant testing everything completely before deployment at sea. Keith adds: "We have a mantra—twiddle, test and toss [into the ocean]. Twiddle means to tinker with something until it works, so you'd work on a piece or component, then you test it, and finally you toss it in the ocean. You can't twiddle and then toss, because you're going to have problems. The testing part is crucial. So we set up a program at UVic for receiving all the cabled observatory components from a variety of vendors, checking everything was correct, then testing it in a saltwater tank. As a result, when we put it on the bottom, we knew it would work. And it did." Highland achieved a very high success with 90 per cent of the equipment working and reporting back to shore when first switched on."

Keith's company also used ROPOS II to latch onto the instrument frames or cable drums (some weighing up to 4,000 lb/1,800 kg), launch over the side

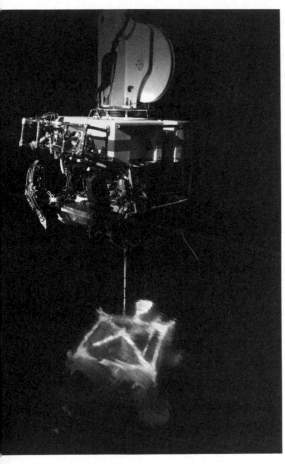

The ROV *ROPOS* safely lowers a subsea observatory module into the water. Personal collection of Verena Tennecliffe

of the ship, descend to the seabed, deploy the equipment within 1 m (3 ft) of the planned location, unlatch and then continue the dive as a free-swimming ROV. The result was a considerable saving of ship time and ROV time. "We put this huge *NEPTUNE* observatory in the water in one year. So it was disheartening when the University of Victoria's Ocean Networks Canada took over, because they preferred to bring in ROV systems from the US that have less reliability and capability, poorer navigation and greater cost.

"It's hard to fight against that kind of thinking—that Canada can't do it or that bigger companies or countries are better. That affects all of our livelihoods." Keith cites the recent Canadian helicopter crash in the Mediterranean as another example: "We were perfectly placed to go in and do a search and recovery. We've got the experience and expertise and that depth was no problem. Instead, DND hired an American vehicle to do the recovery. We can do the work better and faster, with greater reliability."

In the last several years, the business model for CSSF has morphed as Canadian marine science has declined. "More people are doing data analysis of the cabled observatory information and fewer people are going offshore to do the physical analysis, particularly in Canada," Keith states. "Another huge problem for Canada is our lack of ships for research. The mothership for scientific research was the *Hudson*, but it's old and unreliable. There's no ship on the East Coast that can handle scientific work, although they're trying to put something together with a charter ship. In fact, our 'newest' ship, the *John P. Tully*, was built in 1985. By comparison, the Dutch have *Pelagia*, which is only 22 years old, and they're already planning its replacement."

The story of the CSSF and Highland Technologies is a familiar one, echoed by many small subsea tech companies on Vancouver Island, in the province and in Canada. "We need institutions like the University of Victoria to nurture companies like ours, not compete. And we especially need recognition from our own government. They need to understand that local companies—and there are a bunch of us—do what we do effectively and with less cost than bringing in outsiders. We need that acknowledgement in terms of contracts. That would have tremendous spinoff in the province and in the country for the subsea industry."

Dr. John Bird, DREP and DREO

John Bird's attraction to the water began on the Prairies, strangely enough. "We had one television channel and *Sea Hunt* was my favourite show! Because my dad was in the Forces, we moved all over, so I grew up across Canada and ended up in BC for high school." John went to BCIT and graduated as a technologist. Because of *Sea Hunt*, he took a scuba-diving course when he turned 16, and was soon teaching it at the YMCA.

During his last year at BCIT, employers came to interview students for jobs, and one of the employers was the Defence Research Establishment Pacific (DREP), as it was called back then. "I did the interview, my marks were good, and the fact that I was a diver helped because DREP did underwater acoustics, so they hired me. I was a technician for them for a year and spent quite a bit of time at sea.

"During the Cold War, a lot of acoustic research was spurred by the fact that radio waves can't penetrate the ocean, so acoustics was the only reliable way to detect stealthy submarines that patrolled the oceans. In fact, most of what we've learned about underwater acoustics is because of

Dr. John Bird has pursued underwater research working in both the Canadian military and at Simon Fraser University. The challenge, he notes, is that putting equipment underwater is never as easy as using it on land. Personal collection of John Bird

military objectives. People can say it was science, but in fact it was the military objectives that produced the science. Those submarines *were* out there and had to be detected. I remember one sea trial with DREP when we were doing research towing a long array of sensors behind a ship. We had another ship with an acoustic projector that we pretended was a submarine and we tried to detect it with our array. All of a sudden, we realized that we were not alone. There *was* a real unidentified sub out there! It just made the hair on the back of your neck stand up."

At the end of his first year as a technician with DREP, John's boss said, "The kind of work you're doing is science; maybe you should go back to school." His wife agreed to work so long as he could get scholarships. "That's how I ended up at UBC in their engineering program. After graduating, I went back to DREP, this time as a scientist."

John continued working for the military for three years, running various programs at sea, where there were always challenging technical glitches and weather. He tells of equipment that ended up "permanently deployed"— the euphemism for equipment lost at sea and likely non-recoverable for one reason or another. "The most dangerous time is when you're putting equipment in the water or retrieving it. The ship is rolling, so you've got all kinds of dynamics going on. That's when deployment and retrieval can become a nightmare. We lost one set of equipment when the steel cable snapped; that happened because it was corroded by a little bit of an electricity leak in saltwater. The next time, we used a Kevlar cable, but it chafed as it ran over a sheave [the groove in a large pulley], so we lost that set of gear, too."

Another particularly dramatic event happened when one of the towed arrays was running off the huge hydraulic-powered winch at the stern of the ship. All of the lab gear was in a dry lab close by the winch. John recalls, "I turned on a piece of equipment in the lab and suddenly the lab went black because there was too much of a current draw. But the real problem was that the hydraulic pumps that powered the winch were on the same circuit as the lab. So here's this big chunk of equipment out there in the water and now it's dragging the tow cable off the winch. I could see this big drum on the winch, that's twice the size of a person, just slowly starting to spiral out cable. It was turning very slowly in the beginning, but picking up speed. The mechanical brake on the winch was smoking and ineffective.

"I'm thinking, 'This isn't going to end well.' So I ordered everybody off the quarter deck, into the lab, and we closed the hatch. I called the bridge and told them to put the ship in full reverse so the cable isn't dragging out so fast. They tried to do that and tried to get the hydraulic pumps going again, but we hit the end of the cable first. Wood blocks at the centre of the winch's drum flew all over the place. Inside the lab we could hear the banging and crashing. Then suddenly everything was quiet. I went outside and here's this big empty drum just rocking slowly back and forth with the motion of the ship. The big A-frame at the stern that the cable ran out over now had a huge bend in it. We were just fortunate that we got out of the way in time while our towed array was being unintentionally 'permanently deployed.'"

Eventually John tired of spending so much time at sea, away from home. At that time there was a DND program that allowed a return to the school of your choice and paid tuition, moving expenses and even three-quarters salary. "It was perfect! I asked to go to Carleton, in Ottawa—my wife wanted to skate on the Rideau Canal." The program gave John three short years to go from a bachelor's degree to a PhD. The first year he had to write the PhD comprehensive exams before they'd let him into the PhD program; he also took 13 courses. "I also started on my PhD thesis and had it ready to defend in two years and eight months." Then it was back to Victoria and DREP.

Since he'd been funded for three years of studies and was required to work two years for every year he'd been away at school, John was committed for the next six years. However, house prices in Victoria were going up like crazy, so he asked about a transfer from DREP to DREO (Defence Research Establishment Ottawa). Given the OK, John and his wife moved back to Ottawa, where he worked on space-based radar and satellite communications. "Near the end of my six years, a headhunter contacted me because the University of Saskatchewan was looking for someone like me for an academic position. I hadn't been thinking of that, but I flew out for the interview. Unfortunately it was in the middle of February, and I knew I wasn't interested in somewhere that was even colder than Ottawa."

SFU and the Underwater Research Lab

In the process of getting references for that application, John had contacted his former PhD supervisor, who was now at Simon Fraser University as the Dean of Applied Sciences and tasked with starting an engineering school. He invited John out to talk. "That's how I ended up at SFU in 1987. He was very interested in subsea things, so we brainstormed the idea of an engineering science lab focused on the ocean—and we called it the Underwater Research Lab. Eventually we hired Harry Bohm as lab manager."

Subsea work requires a generalist mentality. That's because putting equipment underwater is not as easy as using it on land. "You have to be competent in many fields, from navigation to how to waterproof equipment, to the strength of materials, to the motion dynamics of structures—everything! So you really have to have an intrinsic interest in the oceans because there's just so much work involved in putting things underwater."

Dr. John Bird is proud of the research collaborations that came out of the lab before he retired in 2014. For years, he spent one day a week with International Submarine Engineering. In return, ISE helped fund student work in the lab. "We had undergrads working on projects, Master's and PhD students working on their research, visiting scientists and even industrial people on board from time to time. Some of my students have gone on to form their own sonar companies, one works for a drone company and another works for ASL Environmental Sciences. One PhD student's sonar project was patented and became a commercial product." It adds up to a pretty impressive legacy.

Glen Dennison, citizen scientist

Glen Dennison is a compact, positive guy in his early 60s, with an easy, radiant smile. He's been described as a "citizen scientist," yet he has almost no professional affiliation with ocean studies. Instead, he's a self-taught underwater expert who's respected by scientists, DFO researchers, and students in engineering, physics and oceanography. He's also co-authored and presented a number of scientific papers, as well as having written a book on the unique waters of Howe Sound. He funds all of his research out of his own pocket and has built much of the equipment needed, like specialized cameras (drop cams), in his home workshop.

Professionally, Glen is an electronics technologist, recently retired from TRIUMF, Canada's particle accelerator centre at UBC. Now he's affiliated with DFO's Pacific Science Enterprise Centre (PSEC) in West Vancouver, where he has his own lab for marine research in Howe Sound.

Glen started scuba diving when he was just under 16: "I needed a note from my mom!" His diving certification was with the National Association of Scuba Diving Schools (NASDS). "Half of the class failed the exam, but I came in second—that's how keen I was to learn to dive!" Every week, Glen would also hike over to his grandpa's house to watch *Sea Hunt* on TV.

He got his electronics degree at BCIT, taking options in controls and telecommunications. After school, there were stints at International Submarine Engineering (ISE) and with Fairchild Semi-Conductors in California. In 1980 he got a great dry-land job at TRIUMF. But his love of scuba diving is what consumes most of his free time, particularly dives exploring Howe Sound.

Howe Sound is the contemporary playground of thousands of boaters and anglers, but most BC residents know only its surface waters. Glen knows the Sound intimately from below. Most particularly, he's the person who discovered and documented most of Howe Sound's ancient glass sponge reefs. That's no small accomplishment, given how long they were undiscovered: the Howe Sound reefs likely started 6,000–12,000 years ago—possibly right back to the time of retreating ice caps in British Columbia.

Glen enthusiastically explains that glass sponges can be as large as a small car. They exist in an almost sunless world of intense pressure and cold, and they grow on bioherms, mounds of the accumulated fossilized remains of

Glen Dennison has logged over 3,000 scuba dives, many in Howe Sound, in order to document its ancient glass sponge reefs. Photo of Dennison and glass sponges by Adam Taylor, from personal collection of Glen Dennison

previous generations of sponges. "They are remarkable filter feeders, capable of filtering 900 times their body volume per *hour*! These marine 'animals' are actually multinucleated cells." He adds that they have no brain, no nervous system as we know it, and no single gut, but rather many small channels that digest bacteria in their tissue, making the sponge an animal of many digestive guts. In a rare sighting, Glen observed one sponge "clean" itself by coughing or back flushing. "I've probably spent more time than any human observing a sponge bioherm, but I've still only seen that happen once. It was a very dramatic action—not subtle at all."

Having logged nearly 3,000 dives, Glen has used a depth sounder, digital compass, chart plotter and GPS, as well as an underwater closed-circuit camera, to create three-dimensional colour bathymetric maps of more than 40 dive sites in Howe Sound. That material forms the bulk of his book *Diving Howe Sound Reefs and Islands*. These maps also serve as the template for a network of voluntary "no-take" marine sanctuaries. Proceeds from the book helped deploy mooring buoys so that boat anchors do not damage the fragile beauty of the glass sponge reefs below.

Glen works with four highly experienced teams of divers—the UBC AquaSoc, Burnaby's Ocean Quest team, the Marine Life Sanctuaries Society (MLSS) photo team, and a deep-diving tri-mix team, all equally committed to researching, documenting and preserving these ancient, fragile life-forms. To date, they are working on several projects:

- A sponge larvae study involves learning how to repair and regrow glass sponges, particularly in Haida Gwaii.
- A project monitoring temperature on various bioherms, which are the heaps of dead sponges on which live sponges grow.
- The task of taking 300–400 pictures of a single glass sponge in order to create a 3-D image of it.
- Discovering and protecting previously unknown glass sponge reef sites—overall the most important project.

In addition to thousands of scuba dives, Glen has also done scientific observation aboard the submersible *Stingray 500*, operated by Vancouver-based Aquatica Submaries. And, along with Dr. Chris Harvey Clark, he even set the sub's endurance record of 10.5 hours, on a night dive studying resident six-gill sharks. As well, he and scientist/science reporter Alan Nursall got their longest-ever observation of glass sponge reefs in that submersible.

Working with the DFO, various stakeholders, Vancouver Aquarium researchers, First Nations and scientists, Glen was delighted when Fisheries announced the creation in 2019 of eight new marine refuges in waters northwest of Vancouver in order to protect newly discovered glass sponge reefs. The announcement noted that these reefs are also some of BC's most biologically productive areas, providing habitat for more than 84 species, including prawns and rockfish.

"If people realize what's down there and how fragile it is," Glen says, "they tend to get more cognizant of what not to do—so no fishing and no crab or prawn traps in those areas. And they should always report poachers to the Fishing Violation phone number."

The bioherms on which glass sponges grow in the Strait of Georgia are pretty much wrecked, Glen notes sadly. That's why he's committed to preserving Howe Sound, which has several high pinnacle reefs that are particularly vulnerable. "What we do know now is that these glass sponge reefs are not very common in the world and that they are very fragile." To that end, Glen heads up the Marine Life Sanctuaries Society, a group determined to identify and protect as much of the sensitive BC underwater environment as possible. They also hope to establish underwater parks, similar to surface parks, and to reduce ocean pollution.

Glen Dennison is a fine example of what can be achieved by using basic underwater technology and working tirelessly with groups of motivated divers and stakeholders. As for curiosity and commitment, this citizen scientist is indeed remarkable!

CHAPTER 14

Getting into the subsea business today

The first two years I spent interviewing people for this book was an eye-opener. I'd already earned my sea legs on tugs and dredges, fishing boats and tankers, water taxis and cruise ships, writing about the work their crews did. So I knew a fair bit about working *on* the water. But it was especially fascinating to talk with those in the subsea industry, learning about their unusual work, what they'd achieved, the risks they'd taken, the failures that sometimes led to successes. But, I wondered, what about now?

Then I interviewed Erika Bergman and Alison Proctor, two young BC women who are involved in today's subsea world as submersible and AUV pilots. I also talked with brothers Ben and Daniel Hurd, who initially travelled diverse career routes before both coming back to Atlantis Submarines, the dream business their father started (see Chapter 10). As for financial risk and commitment, there is Scott Waters, a young man from landlocked Kansas with the unlikely dream of buying a vintage submersible, refitting it and putting it back to work.

These young people can easily trade stories, aspirations and determination with any of the other 40-odd vintage innovators I interviewed. Young people *are* involved, I realized. You just have to look for them. And the same is true of subsea businesses today—each interview I did suggested more contacts, more companies, more pages in a book that was already pushing its assigned limits. My hope is that reading the first part of this chapter might encourage interesting career considerations—and maybe even crazy dreams. Welcome to today's underwater world!

For the second part of this chapter, for every interview I conducted, I inevitably wondered—and asked, "What were you like as a kid? How did you get started?" And then, because I can vividly remember my own sons struggling with their career aspirations, I also asked, "What advice would you give a young person thinking of going into your field?" Many of their suggestions shape the second half of this chapter. It's good advice, especially if you're at all curious about the exciting challenge of watery work.

Erika Bergman, submersible pilot

Having just turned 30, Erika Bergman seems like the quintessential young person of the future. She's equally comfortable whether doing safety checks on the *Stingray 500* submersible in West Vancouver or narrating a two-hour live broadcast for the Discovery Channel while piloting the submersible in the depths of Belize's "Great Blue Hole." She was also named a National Geographic Explorer in 2013 for her "Classrooms Under the Sea" expedition, which featured live streaming video from submersibles.

Erika has "flown" more than six different submersibles, but her experience is backed with academic credentials, with a double major in chemistry and oceanography from the University of Washington. While studying, she also gained credibility working on the engines of two vintage ships—the *Virginia v* and the *Lady Washington*. Then Erika joined Aquatica Submarines as chief pilot, operations manager and pilot instructor of the company's three-person submersible. "Half of the time I was in the shop doing submarine maintenance or in the water. The rest of it was spent in the office, organizing operations and setting up expeditions around the world, then pitching those ideas to National Geographic, Discovery and other media organizations.

"Piloting submersibles is great, but the joke is that we're just the bus drivers. In fact, it's a biology or geology lesson every time you take a scientist underwater. For example, on board the sub, they not only see the evidence of the current they predicted, but they're actually part of that current." So science is a driving factor for the *Stingray 500*, but it's a very costly one.

Subs are expensive to operate, Erika explains: "So if we're not on a filming expedition funded by Nat Geo or Discovery, then dive days get sponsored

When flying the *Stingray 500* submersible, Erika Bergman worked closely with Crosland "Cros" Seville (centre) and Bodhi Wade (left). Personal collection of Erika Bergman

by what I call philanthropic adventure tours." That's the term for enthusiastic individuals with deep pockets who want to spend their money on something different and interesting—but they also like to see how their money might help out. So they might pay for an entire three-dive day but only go out on one of the dives themselves. The other two dives go to scientists or researchers who couldn't otherwise afford those underwater expeditions.

In the Aquatica shop, Erika works closely with Crosland "Cros" Seville, senior technician and chief boat pilot, and Bodhi Wade, who recently joined the company as pilot-in-training and media coordinator. "We're a very tight-knit group and everybody plays multiple roles." A typical dive-day for the threesome starts early. "We do all the pre-dive checks, working as a team to get everything set up on the sub and with the surface boat. Then the clients show up and we get them aboard the sub and head out on the surface, following the boat. Often, we take them down to a glass sponge reef that's local and quite remarkable. I think we've done almost a hundred dives to that site, mapping the edges, so we've taken DFO and Vancouver Aquarium folks down there and tourists on discovery dives. After an hour or two, we transit back into the hoist and pull the submarine out of the water. They take their data or photos and wave goodbye. Then we do all the post-dive checks on the submarine, following another set of checklists to close it down for the night. The length of the dive doesn't really matter—the entire procedure pretty much takes the whole day."

Erika and Cros are both certified pilots for *Stingray 500* and can handle all the surface support as well, so on any given dive day they rotate jobs. While Erika's background is academic, Cros initially gained experience in construction and working outside. He also did a commercial diving course in Kelowna with Diving Dynamics, and then completed a welding engineering technology program at the Southern Alberta Institute of Technology in Calgary. In between, he spent a year and a half working as a boat captain and dive instructor in the Grand Cayman Islands. "My commercial diving background, a passion for the water, the technical training and always being very mechanical growing up was kind of the perfect combination for what I'm doing now."

Bhodi is just into his 20s so he hasn't had time to amass much work experience, but he was smart enough to begin by doing odd jobs at Aquatica, indicating his interest. That morphed into a full-time job offer. "When I first started, everyone called me the senior optical technician because one of my main jobs was to polish the sub's bubble dome. It was a bit of a running joke but it gave me a fun, impressive-sounding line to put on my resumé, for sure."

Erika notes, "With a small company like this there's a place for lots of different skills *and* being a submarine pilot. I always regretted not being a mechanical engineer. Heck, I didn't realize it was even possible to take those courses, so I did other applicable studies, but there's room for every skill." Everyone agrees that it's important to put one good line on your resumé every year, something that says you're interested in this kind of work whether it's diving or studies or even building things.

Diving submersibles in BC's Lower Mainland waters is considerably different from tropical waters with unlimited visibility, Erika notes. "It's good training to be diving here on the glass sponge reefs. You have to rely more on your instruments, and it's wise to have those technical skills learned in a difficult environment. When you have 100 feet of visibility in the tropics, you don't rely on instruments quite as much."

Submersibles with human passengers always rate higher on the risk scale, and Erika recalls her pilot training, particularly the module on risk assessment. "We all read the book *No Time on Our Side*, about the harrowing rescue of the two men aboard the *Pisces III* that sank off the coast of Ireland. So many people were put at extreme risk to save these two guys, and thank God, they did save them, but it made for an interesting discussion about assessing risk. As a result of reading that book, our policy is that we always fill up our scrubber [air cleaning] cartridge after each dive, even though it's rated for several dives and it costs $20 more. It's just good insurance. Also, if somehow something doesn't feel right in your gut, that's what you have to listen to before you descend."

Erika acknowledges how few women have jobs in any area of subsea work: "I've had the opportunity to live an extraordinary life. But I dream of having other female collaborators, so that's why I started running girls' underwater camps. Most of the participants are either high-risk girls or those from third-world countries. I think I've reached about 500 girls in camps of 20 kids each. My goal is to bring these girls along educationally so that eventually I can hire them! Of course, not everyone will go on in this field, but at least they've had a glimpse of what they might do. And two have graduated from a technical high school now, where they became certified rescue divers as well. One of them is heading off to the US Naval Academy and one is going to Woods Hole. That's all coming from the first moment when they arrived at Girls Underwater Robot Camp in Tampa, Florida, as youth at risk. I'm really proud of them!"

Erika, Cros and Bodhi talk about the future, particularly the need to map the oceans. But as Erika notes, "Mapped and conserved are two quite different things. Four per cent of the oceans have been mapped, but only one per cent of the oceans has been protected. There's scientific evidence that if you protect 50 per cent of the ocean, then the fisheries outside of that protected area will still be lucrative and sustainable. So I think it *is* possible to conserve huge chunks of the ocean, but it's going to require young people committed to moving that agenda forward, whether they're doing it in the field of oceanography or in political office. It's all about awareness."

Erika has also run girls' underwater camps, focusing on participants who are either high-risk or from third-world countries. Courtesy Erika Bergman

In 2019, Erika was faced with a very different future consideration when Aquatica unexpectedly closed its doors—a shock for everyone. "I'm fine, personally, because I have 10 years of experience and wide connections in the subsea world, plus I like doing international work. So it wasn't long before I got a call from a guy needing a private industry sub pilot for work in Florida and the Bahamas." Erika is also pitching a number of ideas to National Geographic, doing a lot of public speaking and even thinking about writing a book. She is also eager to refresh her non-profit society Geeks.com and get back to work doing underwater camps for girls. Philosophically, she concludes, "Working in the underwater world requires a lot of different skills. So does surviving in this business. You have to combine a lot of factors and skills to stay afloat; it's all part of the interesting challenge of this work."

Alison Proctor: AUV pilot and PhD

Alison Proctor has always loved the water, having been raised on Salt Spring Island, taking swimming lessons as a kid, becoming a life guard, doing scuba training and then becoming a scuba instructor aboard cruise ships. But she also loved education and decided to go back to university at Embry-Riddle in Florida, one of the top aviation and aerospace schools in America. Earning a degree in aerospace engineering, she then took a master's degree, specializing in flight dynamics and control systems, at Georgia Institute of Technology.

Determined to continue with schooling, she decided to return to Canada. Her options were Memorial University, where she could continue with her work in aerospace robotics, or the University of Victoria, where she would need to switch to underwater vehicles. "It was a tough choice, but ultimately, my goal was to come back to Canada *and* be close to my family; Newfoundland is not close." Alison met Colin Bradley in UVic's engineering department, a professor who would have a profound effect on her career. During the nine years she was in Victoria, she not only earned her PhD, but learned to pilot and help maintain and adapt a Saab *Seaeye Falcon*. "Learning to pilot an ROV just takes hours of practice, but learning the maintenance and building little tooling packages to facilitate various underwater missions was more challenging. We all had some technical background—we knew the concept of ROVs, telemetry, RS232, network communications and microcontrollers, so it was fun and interesting for all of us."

Alison also helped develop an AUV research program at the University of Victoria. It was a time when the *VENUS* and *NEPTUNE* underwater observatories were being installed, and there was a project to build an ocean technology testbed that could facilitate assessing new sensors and electronics for those observatories. Doing that also enabled Alison to set up and operate a tracking arena for acoustic and vehicle research. A successful grant application allowed UVic to purchase a *Blue Fin* autonomous underwater vehicle (AUV) and an array of sensors for it from local BC providers. "AUVs weren't really new at the time, but oceanographers weren't totally trustful of them. They wanted to see that AUVs could get data that matched what they'd been getting from more traditional methods. So that's what we took on. And it allowed instrument manufacturers to work with us and prove that their previously stationary product could work on a moving vehicle like an AUV."

Alison feels strongly that academic research and development is vital: "R&D happens in commercial settings, for sure, but it's always mission-driven. It's also heavily weighted towards 'safe' R&D where you know there's going to

After getting her PhD, Alison Proctor joined the Burnaby-based company Ocean Floor Geophysics. One of her main responsibilities is piloting the large AUV *Hugin*, shown ready to launch. Personal collection of Alison Proctor

be a profitable outcome at the end. Academic research allows some 'blue-sky' thinking and being able to do things that aren't necessarily going to generate a profit, but will advance the technology. Testing new equipment in the field of engineering always takes money—you have to design the piece, build it, test it and prove it. The same is true in the academic environment, but you also need a project leader who has the right credentials and contacts. As well, you have to be able to write a successful grant proposal. And competition for funding is getting tougher every year."

In 2014, Alison earned a PhD for her work on autonomous and remotely piloted underwater vehicles. Laughing, she adds, "I've definitely spent a lot of time in school!" But that paid off when it came time to look for a job. Her top two options were MBARI (Monterey Bay Aquarium Research Institute) in California and Ocean Floor Geophysics, a business based in Burnaby. She opted to work for the commercial company and feels she definitely made the right choice. "Ocean Floor Geophysics is a really incredible company to work for. It's filled with PhD-level people and others who may not have that degree but have equivalent experience and wisdom. It's very research-oriented and they take on interesting data acquisition, analysis and exploration projects, so you get to exercise your brain."

Initially many of the projects she worked on were ROV-based. Then a few years back the company purchased the large *Hugin* AUV, built by Kongsberg in Norway. Alison serves as the AUV's remote pilot, working from shipboard, and its program manager. "Because we're a small company, when *Hugin* goes out, I go out." She adds, "When you're working at sea, it's just not the same as working at home. One of my co-workers always says that when you're on a ship you might as well be on the moon. That's because all you have is what's around you and your wits to get through whatever problem comes up. In addition, you're always dealing with a formidable foe. Mother Nature and the ocean are trying to get in your way whenever possible. There's never a dull moment.

"Also, there's the reality of the time factor at sea. Downtime is big money, so if you've got a problem with an AUV that should be running a survey, the company wants it back in the water and working, which means *you* need to figure out how to fix it so the AUV can get back on the job." That problem is not unique to the subsea industry. Any industry with huge cash flow, like major airlines, or the film industry, or the oil and gas business, knows the effect of stopping operations or production even for as little as ten minutes. The result is huge pressure on engineers and workers to get a move on, figure out the problem and fix it."

Alison recalls a UVic co-op student who had just come back from working in the Alberta oilfields: "He told me that the most important thing he learned there was how to say no. That's a really hard and valuable lesson. When everybody is demanding that you get the system back up and running and it's not ready, it's hard to stand in front of them and say no. Fortunately, these days if you think that it's not safe to work, people respect that and don't push you. And it comes down to contractual obligations—if you're broken down for 24 hours, then the client gets compensation for that. But for sure, it takes a bit to get your head wrapped around a hundred thousand dollars out the window or half a million dollars gone because you can't quite figure out the problem."

Although there's no question that Alison Proctor loves her job, she now has to balance family life with time at sea. She and her partner have a young son, Winston. "When Kelly and I started a family, we knew what my job

Alison Proctor, shown with her son, Winston, is strongly committed to her career and to her young family. She credits her partner for making that work. "We both work hard whether it's at home or at sea." Personal collection of Alison Proctor

entailed, but that doesn't make it any easier when I have to leave my son. It's a lot of work for Kelly when I'm gone, but we both acknowledge that we are each working hard whether it's at home or at sea. It also requires communication and respect for each other. Fortunately, most of the places I go off to have good VSAT coverage. My son, Winston, figured out pretty quickly that the video me and the 3D me were the same person."

As for gender-based issues, Alison says, "I've been extraordinarily lucky in my career that I've always landed in places where no one reacted to my gender whatsoever. Lots of women aren't that fortunate. It's a struggle to find female engineers or deck crew, so often I'm still the only female. But when I'm in Norway, where our AUV is from, the gender balance is equal —half women, half men. Even in a company lunchroom or meeting, it's 50/50. Here, it's not that way at all, especially in any job on the water.

"When I came back to BC, I had no idea about the historic role this province played in subsea engineering and technology developments. Also, I didn't have any idea about how strong we currently are in that field. We actually have a really, really solid subsea sector with instrumentation, vehicles and operators. I think that Ocean Floor Geophysics is part of that. And companies like AML Oceanographic in Sidney, Inuktun in Nanaimo, and Rockland Scientific International in Victoria are all making new advancements and innovations. But hardly any of that is visible to people outside of the industry."

One change that's beginning to happen even now is increased automation. "We recently did a project where there were two autonomous vessels interacting with each other and going out to complete a mission that previously would have taken a ship full of people to do. I was part of an international team competing in the Shell Ocean Discovery XPrize, and in May our team won!"

Alison easily envisions a future where data processing moves off a ship and is done on land. "It creates a bit of a disconnect between the people collecting the data and the people analyzing or processing it. But any ship owner or company will agree that the more people you have on board, the higher the cost and the higher the risk.

"Unfortunately, all this attention to budget still doesn't translate into more awareness and action about the critical role oceans play in climate change. If the amount of money that currently goes into cell-phone technology was applied to underwater positioning, acoustics and surveying the seafloor so that we could finally map it, it would be done. The problem is that for the most part there is no commercial value in it, so it doesn't happen. Somehow we've got to change that."

Scott Waters and his *Pisces VI* team

HYCO's *Pisces VI* set "first-ever" records in the 1970s and '80s, and was perhaps the most notable submersible International Hydrodynamics (HYCO) ever built. Years later, *Pisces VI* was purchased and went to work for International Underwater Contractors, an American diving company. Eventually it went into storage when ROVs became viable alternatives for the oil and gas industry. Then in 2015, this remarkable submersible got a reprieve when 29-year-old Scott Waters of Salina, Kansas, bought the sub from the Galerne family. In his early 20s, Scott had taught himself how to weld and spent five years building a *Kittredge 350* sub. Now he had bigger dreams. While

Scott Waters (5th from left) and his initial crew for the refit of *Pisces VI*. Personal collection of Scott Waters

he couldn't pay anything like the full asking price for *Pisces VI*, the owners agreed to the sale because he presented a serious plan for refitting the sub and putting it back to work.

Scott set up *Pisces VI* Submarine LLC, built a custom shop for the refit, and spread the word, attracting staff like electrical tech River Dolfi, mechanical whiz Carl Boyer, and filmmaker Mike Kaczorowsky, plus crucial volunteers like Vance Bradley (with a background at Perry Oceanographics). "We just kept finding people to come to the team." Asked what's been the toughest part of the four-year refit, Scott laughs and says, "All of it—both physically and mentally. But removing the fibreglass from the main sphere was a real job." As well, the sub's tail sphere had been cannibalized to make into an ROV, and the dropweight was missing. Scott also knew the sub and its command module had to be inexpensive to transport, so the team worked hard to get them to fit into two 20-ft (6-m) shipping/storage containers. That required resizing the sub to be 18 in (46 cm) shorter—quite a challenge.

In the fall of 2019 the team brought *Pisces VI* to North Vancouver for more work and pilot training. "We almost went to Florida, but BC's subsea history was a big draw," Scott says. "We had planned to work with Aquatica, but just as we arrived, we found out that the company had closed its doors." Scott had hoped that Terry Kerby of HURL (Hawaii Undersea Research Lab) might help with pilot training, but the Covid pandemic changed those plans.

At a 2019 meeting at International Submarine Engineering (ISE), Al Trice (left) and Tom Roberts (right) traded insider info, stories and advice about the *Pisces VI* and subs in general with *Pisces VI* Submarine LLC's electrical tech River Dolfi (second from right) and president Scott Waters (second from left). As Tom recounted, "The motto at HYCO was always 'Keep it simple and it will work,' because the minute you complicate things, it will come back and bite you." Photo by Vickie Jensen

As a result, Scott moved *Pisces VI* to Tenerife, in the Canary Islands, and changed his company registry from American to Spanish. The crew are currently working to finish all testing and training and then plan to ship the submersible to Peru. What hasn't changed is Scott's goal for *Pisces VI*—to provide a versatile deepwater platform to meet the needs of the underwater scientific research community. How's that for fulfilling a childhood dream?

Tips from the Trailblazers

Eric Jackson, President of Cellula Robotics Ltd.

"I got a university degree in electrical engineering from UBC, immediately went to work for International Submarine Engineering and stayed for

22 years," Eric says. "ISE gave me a fabulous opportunity to learn and do a wide variety of interesting things. When I left, I did contract R&D management in robotic underground and subsea mining systems while setting up Cellula Robotics Ltd." Today, Cellula has 35 employees and spans three business lines: geo-tech equipment (primarily seafloor drills and support); engineering services (designing and building custom equipment); and building mission-specific AUVs.

"We get a lot of resumés and we do hire fresh university graduates, primarily engineers. It's good to find a school with a co-op program. We hire those students for their co-op term and some end up working for us afterwards. Also, being part of a robotics club or a subsea club looks good on your resumé. Relevant experience certainly helps, but we look for some kind of keenness, for people who are passionate about robotics and/or the ocean.

"BCIT also has very good training programs for engineering technologists. The thing that's good about BCIT is that, for example, if somebody comes out of their electrical program, they actually know how to use a soldering iron! They knew more than just theory and can hit the deck running.

"Subsea robotics work is multi-disciplinary—so there are aspects that are control systems, mechanical, electrical, software, electronic, hydraulic, etc. The subsea field is changing; now there's a lot of work in renewables and in wide-scale mapping of the seafloor."

John Witney, engineer, HYCO and Atlantis Submarines

"The learning in any subsea business is never-ending. And new technology allows us to see things faster and more sensitively than ever before. As well, there's now a crossover with the space industry, with equipment in the sky and in space that allows us to actually look through the oceans and see what the bottom is!

"I took the route of getting an engineering degree—but I learned early that it didn't make me anything special. It just opened some doors. When I joined HYCO after seven years of education, I met guys there with seven years of practical subsea experience. And I used to wonder, 'Who's smarter?' I probably had a better chance of becoming a project engineer because I had a degree, but not because I knew anything more. The degree was just a start. The bottom line was that we all shared a real interest in what we were doing."

Jean-Marc Laframboise, senior technical adviser for
International Submarine Engineering (ISE)

"These days you pretty much have to get some sort of schooling," Jean-Marc says. "But I don't care if you get straight A's, what you need most is to be practical. How does something work? How do you use these things? Today, adults and kids don't know how to use their hands. Hardly anybody has ever changed a tire by themselves. If they've got a flat tire, they just call BCAA. That means we get engineers who can design an ROV, but they don't know the importance of making it easy to be repaired in the field or underwater—that's because they've never had to do that themselves.

"My dad always had a workshop in the basement. I'd tinker with things, take them apart and put them back together. Now most people grow up with video games and no sense of how to use tools. When we interview people here for jobs, I find only one of five might have any practical skill. They're smart enough to pass courses, to get a diploma, but what you use working at a place like ISE might be only a fraction of what you learned in any classroom.

"When I do an interview, I'll say, 'Go sketch this cup for me on the white board.' You'd be amazed that people don't know how to do that. They'll ask, 'Do you have a computer?' And I say, 'No, but here's a pen and the white board. Go to it.' I love the look of shock on their face.

"I also ask them, 'What kind of projects have you worked on? Do you work on cars? Have you built anything?' Sometimes they'll say, 'Oh yeah, we built a robot.' So then I ask, 'What was your role?' 'Oh, I was the program manager. Or I was in charge of organizing the troops.' But who built it? 'Oh, that was other guys.'"

James A.R. McFarlane, *Ventana* ROV pilot for MBARI, ISE

"In this day and age, it's better to have an education than not, but it can be across a wide range of subject matter. I especially encourage young people to learn how to write code. Artificial intelligence, management of data—that's the way of the future.

"When I started at MBARI in 1987, I was the sixth employee. Half a year later, we finished building *Ventana* and I drove down to California to put it in the water. When I got into the office there was like ten database guys there. What's going on here? What does this mean? It meant that David Packard's vision was extraordinarily clear that the biggest task going into the future would be data management. The amount of information collected by an ROV doing oceanographic research is stunning, and it's all got to be time-stamped and tagged. Somehow Packard knew that MBARI would have to manage all the data collected in 4,000 dives over 16,000 hours.

"There is *not* going to be a time when we stop collecting data. In fact, there's going to be a need for more and more people who know how to do that sort of thing, how to read the codes, how to write the drivers. That's the biggest group of employees at International Submarine Engineering. But all that begins with getting real experience whether it's putting Lego together to make something different each time or designing an ROV like the first one I built at age 14."

Mavis Mitchell, subsea designer and BCIT instructor

Mavis grew up as a tomboy. When she finished school she got an interesting but low-paying job as curator/tour guide at the Rossland Gold Mine Museum. One of her brothers, who was in the pipefitting trade at Cominco said, "It's too bad that women can't get paid to learn a trade."

Mavis laughs, recalling how that comment challenged her thinking and changed her life. She had strong math grades so applied for an apprenticeship as a machinist tradesman at Cominco. That's where she got her Red Seal journeyman's certificate, plus some good hands-on experience *and* decent pay. Eventually she went back to school at BCIT in their mechanical technology program, did well, and moved into the mechanical design program. "I got into designing and building subsea equipment and vehicles of all sorts. Now I'm back at BCIT, only now I'm teaching mechanical engineering students.

"I teach the language of drafting—that means that when you put something on a drawing, it provides information to people that know how to read it.

So I teach my students electronic CAD programs such as Inventor and different programming languages engineers use, like Excel and Virtual Basic. I'm also teaching students how to cast metal parts in the machine shop.

"It's sad that there's only the same small number of women in the BCIT program this year as there was when I went through years ago. In the Scandinavian countries, at least half of the graduates in engineering are women. I don't know what's stopping Canadian women. I don't know if it's a perception that engineering is too difficult, or what. I've talked to a few of the girls here, and it's obvious that the girls here like tinkering, just like I did. They've all worked on cars with their dad or their brothers. That's why they're wanting to be engineers. The part I like best is being hands-on. It's a great career!"

Mark Atherton, sonar specialist, Kongsberg Mesotech

"The academic requirements for sonar work are much tougher now than when I got into the business. So here's some realistic advice. To get into a sonar manufacturing business such as Kongsberg Mesotech, you probably need a master's degree for design work.

"Big companies are always buying up the small ones; that means that in the corporate world the governing rule is always profit margin. As a result, there's not as much opportunity to be creative. So the only place you're really going to learn about this business is to be employed by a marine survey company. Start at the bottom with a smaller company. Be prepared to work offshore and spend time away from home—that's how you'll learn the ropes. Work hard and get experience. The moment you stop learning or feel that you're stagnating, it's time to move on.

"Besides your main sphere of interest—electrical engineering, mathematics, materials engineer, computer programming—you need to know how to tie a knot correctly! So get out on the water. Get your Power Squadron ticket or your sport-diving ticket. Those things will provide insight into the underwater world. And they're important because this type of work involves dozens of skill sets.

"Often training time in any company is way too short. So, to be a success in this business, you need to be super inquisitive. That means learning not only how things work, but figuring out why things *don't* work. Be curious. Read. Take books home, open them up and struggle through them. Then get out of the office and see how and where the equipment actually works on the water.

"Lastly, I hope you can discover something in life that inspires you, like I did, something that says, 'This is what I gotta do!' Then get busy and do it."

Phil Nuytten, diver, inventor and owner of Can-Dive, Nuytco Research

"The old hard-hat divers—you know, the guys with the big copper helmets and the air hoses—had a terse way of summing up their trade to aspiring young deepsea divers. They used to say, 'The bullshit ends on the ladder.'

"And they were absolutely right. There comes a moment when the water closes over the top of your helmet and theory becomes reality. The tender says over the phone, 'Are you ready to begin your descent?' And you answer, 'Roger, Topside. Beginning my descent,' even though every cell in your body is screaming, 'Are you out of your mind? Get me out of here! Pull me up!'

"You are now on that ladder. So, what happens next? Well, that's something you will have to figure out yourself. You should have already learned the most important stuff at home, in kindergarten and in grade school, things like: Don't tell lies; clean up after yourself; don't take more than you can eat; don't hurt anyone. What you don't know yet is that those little truths are actually powerful icons of appropriate global behaviour. Ignore them at your peril.

"The most important thing now is to figure out what you want to accomplish. I know that's not easy, but be decisive and do it. Remember that the road of life is paved with flat, dead squirrels who couldn't decide."

Acknowledgements

————

Four years feels like a sizeable chunk of time to work on researching and writing a book. During that time, I was encouraged that Al Trice, now 92 years old, cheerfully answered every one of my phone calls or visits. And Allen "Al" Robinson regularly energized me, saying, "I can't wait until this book gets published. It's about time!" I can only agree.

This book would not have happened without the patience, encouragement, and enthusiasm of everyone I interviewed. But a core group got me started and continued to provide support, background and answers over several years. They include Al Trice, Phil Nuytten, James (Jim) McFarlane Sr. and his son James "A.R." McFarlane, Deloye "Scratch" McDonald, Al Robinson, Jim English, Mike Macdonald, Helmut Lanziner, Willy and Disa Wilhelmsen, John Witney, Les Ashdown and Mark Atherton.

Subsequent interviews painted a broader picture and understanding of the business as the following told me about their work. In alphabetical order, my thanks go to: Erika Bergman, John Bird, Nhat "Jay" Chu, Steve Curnew, Hugh Dasken, Glen Dennison, Doug DeProy, River Dolfi, Pete "Peetie" Edgar, Danny Epp, Gino Gemma, Tom Gilchrist, Doug Huntington, Ben Hurd, Daniel Hurd, Dennis Hurd, Guy Immega, Eric Jackson, Norman Keevil, Terry Kerby, Terry Knight, Gordon Kristensen, Jean-Marc Laframboise, Jim McBeth, Mavis Mitchell, Don Muth, Jeff Patterson, Alison Proctor, Scott Rivers, Chris Roper, Curtis Schmidt, Crosland "Cros" Seville, Keith Shepherd, Captain John Swann, Doug Taylor, Linda Thompson, Verena Tunnicliffe, Glen Viau, Bodhi Wade, Reid Warwick and Alan Whitfield.

As well, the staff and employees at these companies facilitated my interviews, proofreading and the continual quest for images: Atlantis Submarines

(Maria Capili); Can-Dive and Nuytco Research Ltd. (Jeff Heaton, Virginia Cowell, Donnie Reid and and Mara Scali); Canadian Scientific Submersible Facility; Canadian Underwater Vehicles; Cellula Robotics; Horton Maritime; Imagenex; International Submarine Engineering (Lara Smith, Linda McAuley, Steve Nishio, Betty Redman and Rhonda Cairns); Inuktun (Colin Robinson and Jessica Don); Ocean Floor Geophysics; OceanWorks; the *Pisces VI* project; RMS Industrial Controls Ltd.; Robotic Systems International; and T. Thompson Ltd. Disa Wilhelmsen of Imagenex, Virginia Cowell at Nuytco, and River Dolfi on the *Pisces VI* project deserve a special medal.

Occasionally, recollections were hazy, not surprising given that 40 or 50 years might have elapsed since an event. Sometimes there were varying versions of stories, as well as different perspectives, but even that was an interesting challenge. Those I interviewed not only shared their stories and thoughts, but patiently read and re-read various drafts, corrected errors, answered my endless questions, dug up old photographs and provided the names of others to contact. That help was invaluable. Any uncorrected errors are mine.

I wish I had been able to interview every person who was suggested or those I got to know only after the book had assumed its final form. An unexpected delight was the chance to meet some of the next generation of explorers who are carrying on BC's subsea legacy. This book is their book, too.

My only regret is having to cut text, shorten stories or, in some cases, omit them entirely. And there were certainly many photographs and illustrations that couldn't be included. However, much of that extra material will find its way onto the website www.deepdarkanddangerous.ca.

We all have pivotal events that shape our careers. For me it was four years working with Alan Haig-Brown and Peter Robson at *Westcoast Mariner* magazine, an experience that baptized me in BC's coastal waters and taught me the importance of writing maritime history. Along the way I was assisted by institutions such as the Vancouver Maritime Museum, particularly archivist/librarian Lea Edgar, and the City of Vancouver Archives. As well MLA Sam Sullivan and DFO librarian Marilyn Ness helped in finding elusive economic states for subsea technology. Finally, a big hug to my "non-maritime" readers Lynn Rogers, Anneke Rees and Dana Zeller-Alexis, who cautioned me not to get overly technical. As well, Corinne Selby, Stephanie Nesbitt and sons Nels and Luke Powell provided last-minute editorial feedback.

Many of those in the industry have written their own books or manuscripts that I relied on:

- Friend and fellow author Guy Immega shared his research on John Horton and the *Auguste Piccard*, as well as his manuscript "Dream Boat."
- Jim English's hefty, well-photographed memoir *Elevators to Innerspace: An Engineer's Underwater Adventure* was most helpful.
- Al Robinson's memoir *The Ups and Downs of Allen C. Robinson* helped fill in the blanks and provided photos.
- James (Jim) McLaren Sr.'s *We Changed the World* highlighted BC's subsea accomplishments, as did his collection of first-hand recollections that made up *The Canadian Contribution to the No Time On Our Side Story*.
- Glen Dennison gifted me a copy of *Diving Howe Sound Reefs & Islands*.
- Mark Atherton's *Echoes and Images: The Encyclopedia of Side-Scan and Scanning Sonar Operations* and his story of its genesis were captivating.
- I benefited from re-reading Roger Chapman's *No Time On Our Side*, Eric Jamieson's *Tragedy at Second Narrows*, and Tom Henry's vividly told story of *Pisces I* in his book *Westcoasters: Boats that Built BC*.
- Dr. David McGee's historical assessment "Underwater Mobility in Canada, 1800 to 2007," with its detailed footnotes, provided important perspective. It was commissioned by the Canada Science and Technology Museums Corporation, now called Ingenium.
- James Delgado's "On the Waterfront" column in *Harbour and Shipping* magazine provided a great HYCO summary.
- Stephen McGinty sent his just-published book *The Dive: The Untold Story of the World's Deepest Submarine Rescue* on the subject of the *Pisces III* rescue.
- Accompanying freelance writer Anthony "Tony" Davis was a bonus, as he interviewed Phil Nuytten for a July 2021 *Maclean's* magazine feature.

I thank Darren Elliott and Harry Bohm for jump-starting the idea of this book. Harry has been my maritime compass and cheerleader, often joining me on interviews to take photos. My husband, Jay Powell, is my trusted writing critic and loyal adviser. Sharon Nadeem transcribed my initial interviews. Editor Lynne Van Luven helped shape better chapters, and copy-editor Jonathan Dore did his best to ensure they were accurate. A special thanks to Harbour Publishing's legendary team, who believed in the importance of documenting this unknown chapter of BC history.

At the very beginning of this project, in 2017, the BC Arts Council, with the guidance of Walter Kwan, awarded me a grant and then extended the grant period as the project expanded. Such help is crucial to writers at all stages of their career, financially and emotionally.

Index